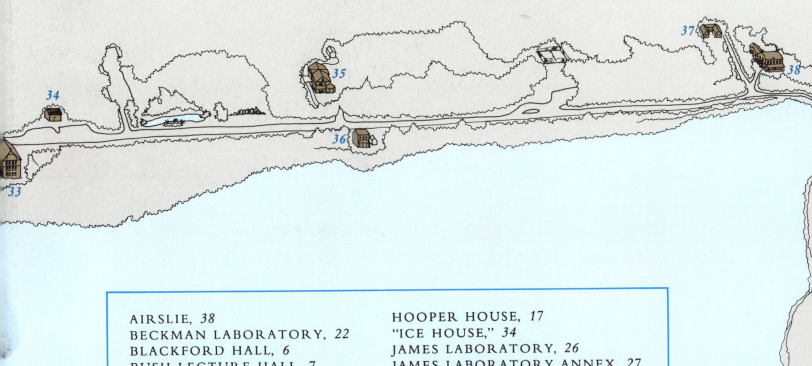

34

35

36

37

38

33

Houses *for* Science

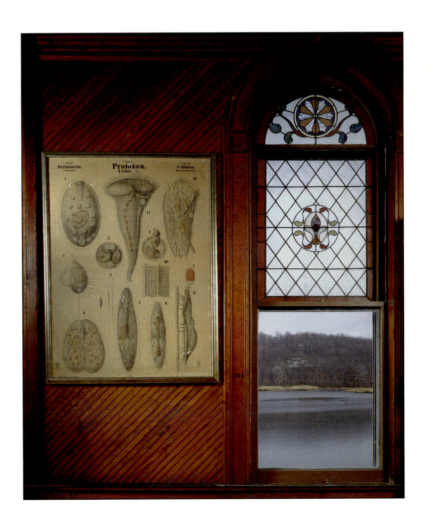

Principal Architectural Photography by Andrew Garn

HOUSES
for
SCIENCE

A PICTORIAL HISTORY
OF COLD SPRING HARBOR
LABORATORY

Elizabeth L. Watson

with

LANDMARKS
IN TWENTIETH CENTURY
GENETICS
a series of essays by James D. Watson

To Barbara,
Best wishes ... and
Happy Bidding at "Bungtown"!
xx Liz
18·V·92

Cold Spring Harbor Laboratory Press 1991

Credits

Andrew Garn i, iii, 3, 12, 15, 26 (*bottom right*), 32, 34, 37, 45, 51, 59, 70, 83, 92 (*bottom*), 94, 97, 99, 104, 108, 123, 132, 133, 142, 147 (*bottom*), 151, 160, 161, 165, 172, 175, 186, 191, 194, 196, 219, 220, 221 (*top*), 226, 229, 237, 238, 240, 244, 247, 250, 251, 252, 254, 257, 261, 267, 270, 275, 277, 281, 282, 285, 316, 317, 319, 322, 334, 335, 336, 337, 338, 339 (*top, middle, bottom left*), 340, 341 (*middle, bottom*), 342 (*top, middle*)

Margot Bennett 29, 163, 198, 245, 249, 299, 312, 329
Mr. & Mrs. Bache Bleecker 8 (*left*), 11
Keith Burridge 223
Timothy Hursley 314, 320, 323, 324, 326
Mort Kelman 302 (*top*), 303
Susan Lauter 144, 183, 233, 302 (*bottom*)
Norman McGrath 202 (*bottom*), 205 (*top*), 206, 213, 214, 215, 288, 289
Peter Menzel 330
Ross Meurer 1, 169, 182, 203, 211, 284, 286, 329
David Micklos 41
Private Collection 130
Jack Richards 109, 111 (*bottom*)
Harvey Weber 24 (*bottom*), 26 (*bottom left*), 58, 66, 110, 112, 128, 129, 131, 140, 155, 180, 190, 207, 216, 224, 339 (*bottom right*), 341 (*top left*), 342 (*bottom*)
Randy Wilfong 306

Avery Architectural and Fine Arts Library, Columbia University, in the City of New York 44
Balthazar Korab, Ltd. 290, 292 (*bottom*), 294, 295
Brooklyn Public Library, Brainard Collection 7 (*bottom*), 10, 13, 33
Centerbrook Architects and Planners 205, 212, 272, 291, 293, 318
Cold Spring Harbor Whaling Museum 8 (*right*)
Cold Spring Harbor Whaling Museum, John D. Hewlett Collection 9 (*bottom*)
Electa Editrice, Milan 292 (*top*)
Frederic Law Olmsted National Historic Site, Brookline, Massachusetts 136–137, 138
Huntington Historical Society 106
Library of Congress 57, 300
New-York Historical Society 24
New York Public Library 65, 73
Peter Aaron/ESTO 270, 273, 274
Society for the Preservation of Long Island Antiquities 86, 87 (*middle*)
Society for the Preservation of Long Island Antiquities, original in possession of Mr. Joseph Ryan 87 (*top left*)
Society for the Preservation of Long Island Antiquities and Vanderbilt Museum 87 (*bottom*)

Illustrations not credited above or in the text are from the Cold Spring Harbor Laboratory Archives or were specifically drawn for this book.

202 © Norman McGrath (1974); 215 © Norman McGrath (1976); 205, 206, 213, 214, 288, 289 © Norman McGrath (1979); 273, 274 © 1987 Peter Aaron/ESTO. All rights reserved; 292 © 1976 by Electa Editrice-Milan (Italy)

Display Photographs

Half-title Page: "Protozoa" wall chart (German origin) inside Jones Laboratory, built in 1893—the oldest laboratory structure at Cold Spring Harbor. (Andrew Garn)
Title Page: Details of the brickwork on Hazen Tower (*left*) and Beckman Laboratory (*right*), both completed in 1991. (Andrew Garn)
Prologue: Reed grass near the head of the harbor. (Ross Meurer)
Chapter One: Cornice detail, 1893 Jones Laboratory. (Andrew Garn)
Chapter Two: Arcade detail, 1914 McClintock Laboratory. (Andrew Garn)
Chapter Three: South gables, 1928 Nichols Building. (Andrew Garn)
Chapter Four: Window bays, 1953 Demerec Laboratory. (Andrew Garn)
Chapter Five: South gable, 1979 Hershey Building. (Andrew Garn)
Chapter Six: East gable, 1987 Page Laboratory. (Andrew Garn)
Chapter Seven: Dormer windows, 1986 Grace Auditorium. (Andrew Garn)
Appendices Part-title Page: Centennial fireworks over Jones Laboratory, July 14, 1990. (Ross Meurer and Margot Bennett)

Houses *for* Science

© 1991 Cold Spring Harbor Laboratory Press
All rights reserved
Printed in the United States of America
Designed by Emily Harste

Library of Congress Cataloging-in-Publication Data

Watson, Elizabeth L.
    Houses for science : a pictorial history of Cold Spring Harbor Laboratory / by Elizabeth L. Watson ; with, Landmarks in twentieth century genetics : a series of essays / by James D. Watson ; principal architectural photography by Andrew Garn.
        p.   cm.
    Includes bibliographical references (p.   ) and index.
    ISBN 0-87969-403-3
    1. Cold Spring Harbor Laboratory--History. I. Watson, James D., 1928- Landmarks in twentieth century genetics. 1991. II. Title.
QH322.C65W37 1991
727'.5'000974725--dc20   91-25921
                                                    CIP

All Cold Spring Harbor Laboratory Press publications may be ordered directly from Cold Spring Harbor Laboratory Press, 10 Skyline Drive, Plainview, New York 11803. Phone: 1-800-843-4388 (continental US and Canada) or 516 349-1930 (all other locations). FAX: (516) 349-1946.

To RVL and JDW with love

CONTENTS

A Village of Science: A Preservationist's View

Cold Spring Harbor Laboratory is the most famous "village of science" in the world, celebrated for its superlative basic research in genetics that now has a 100-year long history behind it. What is equally fascinating is that the "houses for science" in which the researchers live and work on the Laboratory's 100-acre campus on the western shore of Cold Spring Harbor have a history nearly twice as long as that of the science. Many were here long before the Brooklyn Institute of Arts and Sciences founded its Biological Laboratory on this shoreline in 1890. There were homes built for gentlemen farmers and sea captains, as well as dwellings for families engaged in the various industries for which the area was famous in the first half of the nineteenth century. The village of Cold Spring was an active participant in the booming Yankee whaling business, and important supporting roles were played by the mill hands who spun and wove the cloth for the whalers' blankets and shirts and by the coopers who shaped the barrels for shipping out supplies and bringing home the rich bounty of whale oil. Coopering was such a vital aspect of life on the western side of the harbor that the community there became known as "Bungtown," after the "bungs" or tapered plugs used to stopper the barrels. After stoppering, the bungs were trimmed flush with the side of the barrel and one can easily imagine the old north/south thoroughfare that runs parallel to the harbor here—still known today as Bungtown Road—littered with the sawed-off remains of thousands of these stoppers.

There is also a lower portion to Bungtown Road that runs along the water's edge, the main road and the lower one being parallel except for the hairpin curve that connects them (see front endpaper). This connecting bit sweeps down to the water just past the Laboratory's turn-of-the-century dining hall (and also past an old whaling residence that served as the first mess hall back in the 1890s). By the time the road straightens out again it has made almost a complete circle around the most venerable building on the campus, Jones Laboratory, built in 1893 and named after one of the Laboratory's founders. It is the oldest laboratory structure in the United States still in scientific use. As it turned out, this waterfront area was the earliest part of the present Cold Spring Harbor Laboratory campus to be developed, mainly because it had a full range of buildings and structures such as dwellings, warehouses, and wharves that could be readily adapted for use in pursuing science instead of whales.

From "Bungtown" to "DNA Town" is really not such a long story, but it is a bit complex. For most of the Laboratory's first 100 years there were actually two scientific institutions (they are now one) coexisting cheek by jowl along Bungtown Road. The two kinds of science done in the early days were marine biology (and what today would be called ecology), performed under the aegis of the Brooklyn Institute of Arts and Sciences beginning in 1890, and "experimental evolution" (glorified animal breeding and other attempts to prove Darwin's precepts) which was brought to Cold Spring Harbor by the Carnegie Institution of Washington in 1904 at the behest of the then director of the Biological Laboratory, Charles B. Davenport, who soon was wearing two hats. For over twenty-five years Davenport directed both the Biological Laboratory and the Carnegie Institution's Station for Experimental Evolution. When eventually Davenport felt it was time to relinquish the former post he turned it over to his son-in-law Reginald G. Harris, whose claim to fame from the viewpoint of this book is that he managed to get three brand-new laboratories built in as many years, although in appearance they were hard to distinguish from the wood-shingled whaling era houses that stood along Bungtown Road. (The architect of these three laboratories wrote a little book called *Dictionary of Architecture*, published in 1952, from which I have excerpted three dozen entries to form the backbone of the Pictorial Glossary at the back of this book, which I hope will be a boon to neophyte "house nuts.") These Colonial-style laboratories were not nearly as big, nor as solid, as the brick-trimmed stuccoed structures that the Carnegie Institution had erected previously. However, the Biological Laboratory enjoyed the support of its wealthy neighbors on the north shore and so did not feel poor, at least not until the Depression came along. Afterward it was happy enough to join forces in the 1940s with what the Carnegie Institution now called its Department of Genetics at Cold Spring Harbor. This was an informal connection, not an official one, the link being that once again both institutes had the same director, Milislav Demerec. The departure of Demerec twenty years later, however, caused a real institutional crisis. The Carnegie Institution, after erecting a huge (by Bungtown Road standards) concrete laboratory in the early 1950s, could not in the late 1950s find a new director to take Demerec's place and thereupon decided to fold up its tent at Cold Spring Harbor. But how would it unload its buildings here? And what kind of destabilizing effect would this have on the Biological Laboratory?

The answers to these questions and many others can be learned from reading this book, but you can tell a lot just by skimming through the pictures. Yes the Laboratory was saved, but it took the work of a lot of truly dedicated scientists and indeed some mighty talented architects and builders. The architectural firms involved with the various building projects are listed below the respective building heads, which are set in capital letters above the double-columned architectural parts of the text; the firm names are set in italics, together with the date of completion of the project. No listing on the line above the date indicates either that the project was designed in-house by the Laboratory's Buildings and Grounds staff, or in the cases where buildings were acquired long ago that the designer is unknown.

Looking at the Laboratory

by William H. Grover, FAIA

When Charles Moore and I first visited Cold Spring Harbor Laboratory in August of 1973 it looked like a summer cottage village a bit down on its luck. Rampant vines crowded the forests and climbed the trees along Bungtown Road, and the wooden buildings, now so carefully painted and surrounded by manicured lawns, clearly showed their age. Airslie, our first renovation project for the Laboratory, was in sad condition. Every floor sagged and all of the rooms were small and cramped—hardly a welcoming setting for socializing scientists.

The challenge to make good architecture in any situation is made greater here by the very fact that the Laboratory is blessed with a beautiful, albeit at times difficult, setting. We have had to take into account cold springs gushing everywhere, east-facing hillsides that take the morning sun, wonderful buildings (some of which are no less wonderful for being scientifically or mechanically obsolete), residents who care about the Laboratory's effect on their neighborhood, visitors who are impressed (either positively or negatively) by the place, and last but not least the stringent requirements of conducting state-of-the-art science in safe surroundings under the guidance of a director whose concern encompasses all of these aspects. The latter was clearly evident during our first visit as we spent hours divining the exact curve of a driveway that would satisfy Jim Watson's sense of balance and symmetry. We knew this wouldn't be easy, but we had (and still have) great respect for his motivation. He wants it to be right, and so do we.

Jim and Liz Watson have contributed as much as any architect to maintaining the character of Cold Spring Harbor Laboratory while encouraging the improvement and expansion of the scientific facilities. Liz's interest in the history and preservation of buildings, combined with Jim's direction of not only the institution, but also its architecture, has guided our hands (slapping them occasionally) over the last eighteen years. I suspect that a visitor from 1915 would easily recognize many of the buildings at the Laboratory today. Although most have been renovated or added to, they retain much of their original appearance, and the small village character of seventy-five years ago still remains.

There were directions in the architecture at Cold Spring Harbor in the 1950s which, if taken to the extreme, would have caused a much different village to appear on this lovely hillside. Buildings done in the Modernist idiom, such as Demerec Laboratory, would have given the place a cold, institutional look with raw concrete and glass. The Laboratory was lucky, architecturally speaking, to have been unable to afford to build at a time when that style reigned supreme.

During our long association with the Laboratory there have been many discussions about the appropriateness of designs for the various new and renovated buildings. Jim often ponders how to describe the "style" of our efforts, and we have concocted monikers such as "Post-Industrial Collegiate" or "Romantic Pragmatism," but none seem to fit. In our desire to avoid modernism, our work at Cold Spring Harbor might be labeled Post-Modern (as it is on the back endpaper for want of a better label), but we don't employ the usual clichés of Post-Modernist architecture, and we really don't think that the appearance of our work is based on the rejection of anything. In fact, we like to include everything interesting that fits the situation. "Eclectic situationism?" No, that's not it either.

Edward Shils, writing on the subject of "Tradition," said, "Every society derives its identity from its location in time. Traditions are patterns of both persistence and change." Cold Spring Harbor isn't a museum of the past like Colonial Williamsburg. It's really a very modern place, and as architects we believe that to be modern is to learn from and build on history. Architecture that responds to the complex situations of the moment and yet respects the past is the real "modern" architecture. Surveying the changes both scientific and architectural that are described in this book you will also see the patterns of persistence that make the Laboratory such an interesting place to work and live and that establish its identity in this exciting time. As this institution moves into its second century we hope that its architecture will continue to provide variety and interest for its occupants and a sense of tradition to this important place.

We are pleased to have helped build and preserve Cold Spring Harbor as a place that can free the mind to make associations and connections of thought which can result in great discoveries about ourselves and our world.

Thanks to one and all

Cold Spring Harbor—and in fact the world of molecular biology that regularly visits here—surely owes a great debt to the many nonscientists who have built and cared for this "village of science." I would like to single out for special praise Jack Richards, Director of Building and Grounds for the Laboratory, and his incredibly good B&G Department, as well as William Grover, Partner, Centerbrook Architects and Planners, Essex, Connecticut, and his singularly talented design teams. Thanks to them there is a truly beautiful story to tell here. By the way, Jack and Bill have kindly kept an eye on this book through many revisions, helping to eliminate mistakes of the bricks and mortar kind. Bill also contributed the marvellous introductory piece accompanying this Preface, for which I am also most grateful. Jack, as it turned out, is just as remarkably conscientious about using the correct terminology as he is about constructing building details, lucky for us!

Most of the facts reported here about the buildings have been gleaned from the pages of the annual reports and yearbooks of the various scientific institutions in residence during the past 100 years. In fact, they were first presented in the form of a thesis I submitted to the Columbia University School of Architecture and Planning in 1983. Entitled "Science by the Sea: An Investigation of the Architecture and Preservation of Three Biological Laboratories Founded in the Late Nineteenth Century," it compared the buildings and institutional histories of Cold Spring Harbor Laboratory, the Marine Biological Laboratory at Woods Hole, Cape Cod, Massachusetts (founded in 1888), and the Mount Desert Island Biological Laboratory at Salsbury Cove, Mount Desert Island, Maine (founded in 1895 as the Harpswell Laboratory). Visitors to Cold Spring Harbor (and residents too) have seemed to enjoy reading this thesis manuscript on reserve these last few years in the Main Reading Room of our wonderful Library building. What is also stashed away there, but somewhat less accessible, are the Archives of Cold Spring Harbor Laboratory, chock full of ancient printed matter and much, much more, including fascinating old photographic prints. We are all greatly indebted to Susan Cooper, Director of Public Affairs and Libraries, who in 1972 first began assembling a motley collection of institutional records into what has blossomed into a very attractive, eminently usable, and extremely rich archival resource, ably administered today by Genemary Falvey, Library Services head, with the assistance of Lynn Kasso. Susan also read this book in manuscript and although she found plenty of "mistakes" in the draft, her praise was truly inspirational, a boost I certainly needed. Thanks, Susan!

Two others who read the manuscript specifically in the interests of eliminating errors of the architectural/historical kind were Robert B. MacKay, Director of the Society for the Preservation of Long Island Antiquities (SPLIA), and Carol Traynor, Publications Coordinator for SPLIA, Long Island's premier private organization in preservation advocacy and preeminent owner and operator of historic house museums. Actually it was Bob who got me interested in preservation in the first place. In the spring of 1973, fresh from studies in Boston under Professor John Coolidge and newly appointed to head up SPLIA, Bob came by Airslie, the 1806 farmhouse at the north end of Bungtown Road (overlooking the Sand Spit) that for the preceding thirty years had been serving as the official Laboratory director's residence. My husband Jim and I would be moving into this house a half a year or so later, and at the time of Bob's visit most of the old plaster walls had been laid open as the house was being completely replumbed and rewired. How strange it all appeared, at least to my eyes. There was horsehair in the plaster, bark still clung to the studs (after almost 175 years), and strange hieroglyphics were chiseled into the exposed big beam in the hallway that supported all of the second floor joints (these were in fact the Roman numerals essential for accurately pegging the framework of the house together on the ground before it was "raised"). "Bob, where can I learn more?" I asked, little suspecting that this chance encounter would lead to a "career." The rest is history, as they say.

For help in assembling some of the more obscure facts presented in this book, particularly those pertaining to Cold Spring Harbor's illustrious Jones family, I would especially like to thank Bache Bleecker, Ann Crooker, Townsend Knight, and Dudley Stoddard, all direct descendants of Major Thomas Jones and good neighbors and friends. The information on the Joneses is mostly in the early chapters. For help in writing the later chapters, which correspond to the historical period when genetics came of age at Cold Spring Harbor after World War II, I am needless to say greatly in my husband's debt for sharing all kinds of information. He himself was here quite a lot—first as a researcher, soon afterward as an active meeting participant, later as a trustee, and eventually as the director, this last phase being the one that I am very glad to have been sharing with him. Jim also penned the elegant essays on "Landmarks in Twentieth Century Genetics" that grace the beginning of each chapter (and the middle of Chapters Four and Six as well!). His immediate predecessor as director of the Laboratory, John Cairns, was also kind enough to read the book in manuscript form, and his comments, which I have tried to incorporate as best I could, were articulate, brilliant, illuminating, and poignant.

At the Laboratory there are many other staff members who have helped as well. Chapter Seven (the last chapter in the book, devoted to what the Laboratory is today) has seals of approval from all of the following—Morgan Browne, Administrative Director; John Maroney, Assistant Adminstrative Director; Terri Grodzicker, Assistant Director for Academic Affairs; Jan Witkowski, Director, Banbury Center; and David Micklos, Director, DNA Learning Center. The attractive charts for the book, including the building time line on the back endpaper, have been prepared by Michael Ockler at his Hershey Building command post, and Susan Lauter of the DNA Learning Center drew the campus map on the front endpaper. Sue also took several of the color photographs, as did another talented Laboratory designer and photographer, Margot Bennett of the Public Affairs staff. Many of the black and white photographs from the Cold Spring Harbor Laboratory Archives were printed by Herb Parsons, Audiovisual Director. A number of the "postcard perfect" views are by local photographers Ross Meurer and Harvey Weber. (Sadly, Mr. Weber died in early September of 1991 never having seen the beautiful reproduction of his photographs in this book.) The majority of the color photographs were specially commissioned for the book and shot in the course of several energetic photo sessions in the fall of 1990 and the spring of 1991. With Art Brings, Director of Environmental Health and Safety, and staffer Don Rose helping to give the "environment" a boost by turning off lights, adjusting blinds, and shooing away automobiles, principal architectural photographer Andrew Garn, of New York City, did what he knows best, with extraordinary results.

In its overall appearance the beautiful book you hold in your hands is the brainchild of graphic designer Emily Harste, who was lured back to Cold Spring Harbor's shores (after "retiring" from designing all manner of scientific books for the Laboratory) by the prospect of helping give birth to this long-in-gestation work. (The timing was finally right once the Laboratory joyously turned 100, and hard on the heels of the wonderful birthday party the new Neuroscience Center successfully opened to great acclaim; see Chapter Seven for photos.) Emily, thank you for "volunteering" for this monumental task; hope you are at least one-hundredth as pleased as we are, both myself and the denizens of Bungtown Road who have helped flesh out your imaginative design.

This highly skilled bookmaking group ensconced at the top of a hill on Cold Spring Harbor's western shore includes the editorial ombudsman extraordinaire of this book, Nancy Ford, Managing Editor of Cold Spring Harbor Laboratory Press, and her wonderfully capable assistants, Annette Kirk, Production Manager and protagonist par excellence of the photographic print; Dorothy Brown, Senior Technical Editor and guardian of consistency; and Pauline Tanenholz, Office Manager and "keeper of the gate." Visits to Urey Cottage, editorial and production bastion of Cold Spring Harbor's world renowned book publishing enterprises, always started off with writer's chills that quickly turned to thrills as yet another fearsome pack of captions or text were dispatched in Nancy's "den." (She would be the first to confirm that her office is both "a comfortable, usually secluded room" and "a center of secret activity.") Incidentally, it is no secret that this really is Nancy's book. There are no typos, you can be sure of that (although Nancy would caution against making this claim). If there are errors of fact then I am solely to blame. The rest is Nancy's baby; let's toast her, the "birth," and the next 100 years!

ELW

July 1991

Houses *for* Science

PROLOGUE

Before the biologists arrived the western shore of Cold Spring Harbor was called Bungtown and earlier Wawepex

Biologists were not the first to discover the natural abundance of the Cold Spring Harbor area. The colonists who arrived from New England in the early seventeenth century quickly recognized the promise of the area's natural resources and proceeded to establish a variety of industries. For centuries prior to these arrivals bands of Native Americans had encamped at the head of the harbor. Members of the Matinecock tribe of Long Island, they inhabited a naturalist's dream of paradise. Freshwater springs and streams, brackish tidal flats, and a saltwater marsh navigable at high tide were all close at hand and teeming with life. The heavily wooded slopes and hills that sweep down to the harbor abounded with small game. One can easily imagine these earliest settlers enjoying peaceful lives on the banks of the harbor secure in nature's bounty.

ADDRESS OF PRESENTATION

of Wawepex Society Lands to the Carnegie Institution of Washington for the Station for Experimental Evolution
Being a Speech delivered at Cold Spring Harbor, Long Island, New York on June 11, 1904
by W.R.T. Jones, Governor of the Wawepex Society of Cold Spring Harbor

---

Representatives of the Carnegie Institution of Washington, ladies, and gentlemen:

Cold Spring has experienced several distinct changes since Prime, in 1845, wrote his history of Long Island. He devoted to it just four lines, describing it as "a considerable village in the northwest corner of the town [Huntington], lying on a harbor known by the same name." The village had long possessed two factories and a flour-mill, which were of great benefit to the neighboring farmers in taking their wool and grinding their grain; also two or three stores, all doing a small paying business. With the introduction of the whale-fishery business, the village awoke to a real boom. Buildings were erected to accommodate this business, houses built for the employees, and in my early days the village, especially on the west side, showed its activity by noises from the continued hammering of iron, the resounding echo from the coopering shops, the clanging of boat-builders, and the buzzing of saws. When this business became no longer profitable, the place soon appeared like a deserted village—houses became vacant, buildings unused, and everywhere neglect and decay.

The whale-ships ordinarily came to anchor in the outer harbor. My father, John H. Jones, built a dock on the east side of the inner harbor to facilitate their outfitting, and I have seen a vessel fitting out at that dock for a three years' voyage to the Arctic; but the great rise and fall of the tide prevented the experiment being a success, and the original anchorage was resumed. The great rise of the tide—some 7 feet—was in one respect an aid outside, for, lying at anchor several months, the anchors sank so deep in the mud that the windlasses of the vessels could not start them, and when the chains were hauled taut for the vessel to pull by the rise of the tide, it often took several tides before the windlasses could weight anchor, necessitating three days in breaking anchorage.

There were two post-offices by the name of Cold Spring in this State, and the delivery of letters became so confused between the one on the North River and the one on Long Island that the name of the Long Island village was changed to Cold Spring Harbor. It was then made a port of entry, an honor which I believe it still retains, but the income is very limited. Many of the deserted buildings were torn down—one because it interfered with the view of the outer harbor from this house; two or three have been modified so as to be of present use. The inner harbor, with its clear water, was in those days a constant source of amusement. A pretty sandy shore at the lower end of these grounds, with a clean sand-bar extending out, was a delightful place for youngsters, especially from the district school near by, to bathe at medium tide, and I never failed in taking advantage of this sport. A legend was long current that General Washington, on his way from Oyster Bay through the island, halted at this school-house when being erected and gave personal aid in raising the first rafter. At low tide the water largely covered the bottom, and at the deep hole a number of acres were always filled with 5 to 6 feet of water, even at the lowest tide, which permitted a pleasant pastime for young people to fish and secure results worth serving at the table, the incoming tide always bringing in a fresh supply of fish. Occasionally, but at long intervals, one or two porpoises might be seen sporting in the inside water, but as soon as the tide turned to ebb they made for the outer harbor and no effort to stop them ever succeeded, as they dived under or leaped over the string of boats stretched across the narrow entrance to stop their escape.

The next change, particularly on the west side, assumed a scientific aspect.

My brother, John D. Jones, inherited the family homestead and adjoining grounds. He was born in the family mansion, which was destroyed by fire, and he erected this building on the site of the old house. The Brooklyn Institute desiring a place to establish a school of biology, he put up for that institute a building suitable for its purpose, and the school, under charge of able professors, has been a success, doing original work which has been a credit to Long Island, and acknowledged as such by similar foreign institutions. He also leased to the State of New York grounds for a fish hatchery, which is now turning out each year several hundred thousand trout and salmon to stock the inland waters of the State.

Seeing the need of an organization to perpetuate the management and care of the grounds and property devoted by him to scientific research, he incorporated the Wawepex Society under the laws of the State of New York governing scientific societies, and the above society has been in charge for several years. The name is taken from an old Indian name of the harbor. Mr. Jones, one of the incorporators of the society, at its meeting January 25, 1892, to organize, was chosen as governor, and was continued in that office until his death, September 22, 1895.

This year the Carnegie Institution, attracted by the advantages of the locality, has asked for a fifty-years' lease of part of the grounds, taking in this house, for carrying out experiments in evolution, promising to put up a special building for that purpose, and the lease has been granted. It gives great pleasure to the Wawepex Society to pass over to the representatives of the Carnegie Institution the papers putting that institution in possession of as much of the property as it desires for erecting buildings to carry out its experiments. I trust in going back and investigating, as far as possible, the origin and order in creation it will find nothing to interfere with the doctrine of the church just around the corner, erected largely by aid of family relatives, in its efforts for improving morals and explaining to the best of its ability life hereafter.

With these three institutions hailing from our village, it will assuredly soon become well known and appreciated both at home and abroad.

The Wawepex are the first to inhabit Cold Spring Harbor

Native Americans on Long Island customarily lived in large extended family units on or near bodies of water and they would name their band after their locality. The Indian settlement that extended out from the head of the harbor along its western shore was called Wawepex, an Algonquin word meaning "at the good little water-place" (when loosely translated as "at the place of the good spring of water" it is clear why the English settlers would soon be calling the area Cold Spring). Because the Wawepex band lacked a written history (and the colonists from New England never compiled one), little is known about these original inhabitants. Among the few traces left from their long existence here are the countless arrowheads that have been unearthed from one end of Bungtown Road to the other. Most of the present-day campus of Cold Spring Harbor Laboratory is situated along this old road, which begins at the head of the harbor and extends northward along its western shore for half a mile, ending at the Sand Spit. A natural breakwater between the smaller, shallow inner harbor and the broader, deeper outer harbor that opens onto Long Island Sound this narrow spit of land presents two very different aspects: a sandy beach on the north and a grass-strewn marsh on the south. *(1)*

(1) Arrowheads and spearheads found on the grounds of Cold Spring Harbor Laboratory.

This is only a small sampling of the many projectile points found by Laboratory employee Guy Cozza. All of these were made or used by the Native Americans who were the first residents along the shores of the harbor at Cold Spring. The Indian habitation here, lasting from approximately 2500 B.C. until A.D. 1600, comprised several settlements, including an encampment on the high ground to the west of the present-day Laboratory tennis courts. Many of the items depicted here were collected at this site, together with drills, knives, and other items of stone, as well as fragments of clay vessels. According to Cozza, glassy white quartz (beach stone) was the most common material for making arrowheads on Long Island. Items made of various types of flint (including glass-like chert and black flint) are an indication of trade with other Northeast Indian groups.

Ecologically, this estuarine environment was extremely diversified. Aquatic habitats included the seawater of the outer harbor, the brackish water of the marshy inner harbor, and the fresh water of the numerous cold springs in the vicinity, as well as of the Cold Spring River that flowed openly into the harbor. (This small river, known by the Indians as the Nachaquatuck, or "the ending tidal stream," would eventually mark the western boundary of the English settlement of Huntington.) The transition zones between the sea- and freshwater worlds of Cold Spring Harbor would especially attract the interest of the biologists who came later, and this lush world of the Wawepex band would capture the imaginations and hearts of modern naturalists.

Englishmen arrive on Long Island via New England

By the middle of the seventeenth century the Wawepex had yielded their lands to the colonists from New England, who established the two towns bordering Cold Spring Harbor: Oyster Bay founded to the west in 1653 and Huntington founded to the east in 1655. To purchase these lands the town fathers negotiated with sachems, representatives of the Native American inhabitants. Similar land purchases took place up and down Long Island. In time these first purchases became subdivided and farming slowly gained a foothold on the north shore of the island. It was during this era that the ancient river at Cold Spring was dammed to form a series of lakes, and the first grist mills sprung up on their shores. With the abundant water resources and ready access to the sea available at Cold Spring it is not surprising that by the opening years of the nineteenth century farming activities were being supplemented by light industry, in particular textile manufacture.

The Jones family of Cold Spring were descendants of the "pirate" Thomas Jones

The two men most responsible for the newly emerging industry at Cold Spring were John Hewlett Jones (1785–1859) and his brother Walter Restored Jones (1793–1855). The Jones brothers were descended from Major Thomas Jones, a privateer from the British Isles. (The family have denied the appellation of "pirate" for their illustrious progenitor since he carried aboard his vessel a "Letter of Marque." In effect this gave him license to seize vessels belonging to enemies of the Crown and was his reward for fighting on behalf of King James II at the 1690 Battle of the Boyne.) Major Jones arrived at Oyster Bay, via Newport and Jamaica, in 1695. By the time of his death in 1713 he had acquired, in partnership with others, thousands of acres of land from the native inhabitants of what is now the Town of

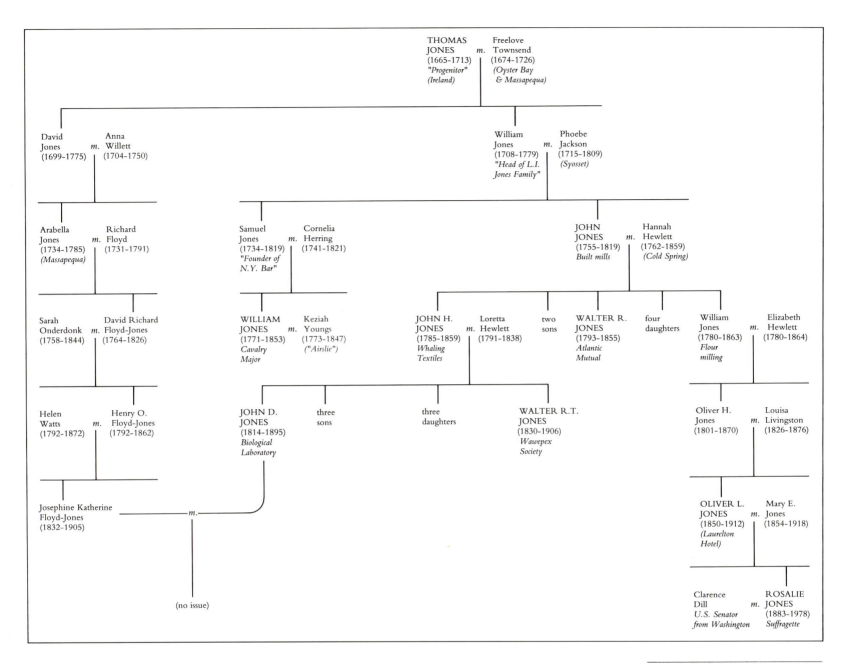

THOMAS
JONES
(1665-1713)
"Progenitor"
(Ireland)

m.

Freelove
Townsend
(1674-1726)
*(Oyster Bay
& Massapequa)*

David
Jones
(1699-1775)

m.

Anna
Willett
(1704-1750)

William
Jones
(1708-1779)
"Head of L.I.
Jones Family"

m.

Phoebe
Jackson
(1715-1809)
(Syosset)

Arabella
Jones
(1734-1785)
(Massapequa)

m.

Richard
Floyd
(1731-1791)

Samuel
Jones
(1734-1819)
"Founder of
N.Y. Bar"

m.

Cornelia
Herring
(1741-1821)

JOHN
JONES
(1755-1819)
Built mills

m.

Hannah
Hewlett
(1762-1859)
(Cold Spring)

Sarah
Onderdonk
(1758-1844)

m.

David Richard
Floyd-Jones
(1764-1826)

WILLIAM
JONES
(1771-1853)
*Cavalry
Major*

m.

Keziah
Youngs
(1773-1847)
("Airslie")

JOHN H.
JONES
(1785-1859)
*Whaling
Textiles*

m.

Loretta
Hewlett
(1791-1838)

two
sons

WALTER R.
JONES
(1793-1855)
*Atlantic
Mutual*

four
daughters

William
Jones
(1780-1863)
*Flour
milling*

m.

Elizabeth
Hewlett
(1780-1864)

Helen
Watts
(1792-1872)

m.

Henry O.
Floyd-Jones
(1792-1862)

JOHN D.
JONES
(1814-1895)
*Biological
Laboratory*

three
sons

three
daughters

WALTER R.T.
JONES
(1830-1906)
*Wawepex
Society*

Oliver H.
Jones
(1801-1870)

m.

Louisa
Livingston
(1826-1876)

Josephine Katherine
Floyd-Jones
(1832-1905)

m.

OLIVER L.
JONES
(1850-1912)
*(Laurelton
Hotel)*

m.

Mary E.
Jones
(1854-1918)

(no issue)

Clarence
Dill
*U.S. Senator
from Washington*

m.

ROSALIE
JONES
(1883-1978)
Suffragette

Oyster Bay. These purchases extended from Long Island's north shore all the way to its south shore—hence the name Jones Beach for that well-known part of Long Island's shoreline. Years later one of Major Jones descendants, John Jones, married into the Hewlett family of Cold Spring, who had been settled in the area long enough to have built a grist mill on the east side of the harbor in 1791 and subsequently a weaving mill (Lower Factory) on the shore of the Second Lake to the south of the harbor. *(2,3,4,5)*

(2) Jones family genealogical chart.

This abbreviated family tree spotlights (CAPITALS) members of the Jones family mentioned in the text. (Adapted from John H. Jones' *The Jones Family of Long Island,* 1907)

(3) Map showing the early eighteenth century domain of Major Thomas Jones.

Progenitor of the Jones family on Long Island, Major Thomas Jones (1665–1713) was a privateer who originally came from the British Isles. Between the time of his arrival at Oyster Bay in 1695 and his death in 1713 at his home in Fort Neck (present-day Massapequa) he amassed thousands of acres on the island from shore to shore, including most of what today comprises the Town of Oyster Bay. He was the "Jones" of "Jones Beach." (Reproduced from *The Jones Family of Long Island* [1907] by John H. Jones.)

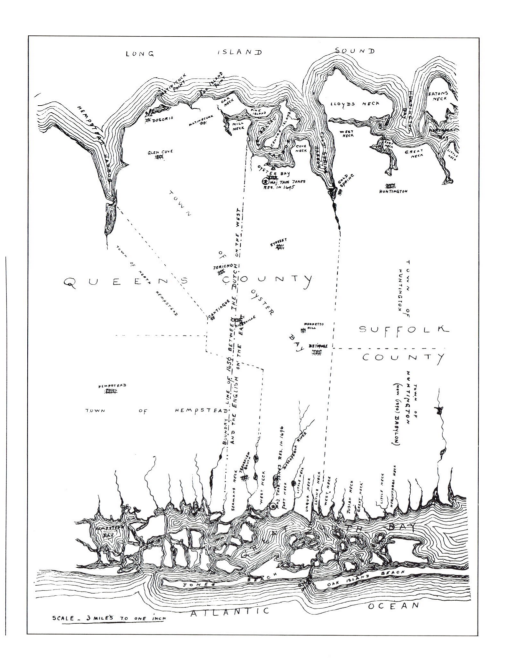

John H. and Walter R. Jones found the "Jones Industries" at Cold Spring

John Jones' sons John H. Jones and Walter R. Jones eventually became part owners of the Hewlett mills and striking out on their own opened a general store on the east side of the harbor. In time they went on to establish such maritime businesses as ship repair and sail making on the eastern shore of the harbor. An additional shipyard, together with a super-

(4) Hewlett-Jones grist mill built in 1791.

Flour milled in Cold Spring was aggressively marketed in Manhattan. The mill was destroyed by fire in 1921.

(5) Lower Factory on the Second Lake circa 1880.

The Lower Factory (weaving mill) was erected at the beginning of the nineteenth century on the shore of the Second Lake south of St. John's Church.

(6) Portraits of John Hewlett Jones (1785–1859) circa 1850 and Walter Restored Jones (1793–1855) circa 1855 by Shepard Alonzo Mount.

The sons of John Jones and Hannah Hewlett, John H. Jones (*left*) and Walter R. Jones (*right*) were the driving force behind the "Jones Industries" at Cold Spring, which by the mid-nineteenth century included flour milling, shipbuilding, textile manufacture, and whaling. The artist, Shepard Alonzo Mount (1804–1868), was the older brother of noted Long Island genre painter William Sidney Mount. Shepard Mount painted portraits of over two dozen members of the Cold Spring Jones family.

 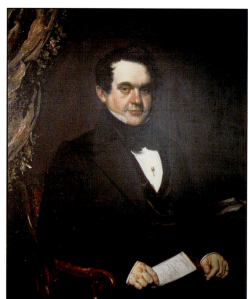

visor's residence, soon appeared on the west side of the harbor. "Jones Industries" is the term that later historians have used to designate all of the various commercial interests of the Jones family in Cold Spring, but in the first third of the nineteenth century the textile industry was king. To increase the output of the weaving mill on the shore of the Second Lake a spinning mill and dye house were erected on the western shore of the harbor near its head. This collection of buildings constituted the Woolen Factory at Cold Spring. (The wool, a high-quality merino type, came from a Southdown breed of sheep raised locally.) A number of multiple-family residences to house workers in the textile factories and their families were also erected, particularly on the west side of the harbor. *(6)*

Bungtown comes from bungs for barrels

A cooper's shop, known as the Barrel Factory, was situated on its own wharf on the west side of the harbor, just south of the ship repair yard. This building housed an activity essential to the milling enterprise carried out on the eastern shore of the harbor—the production of barrels for shipping flour to distant markets. The Jones grist mill garnered grain from a large area, and flour milled in Cold Spring was marketed as far away as New York City. It was from this busy Barrel Factory that the highly commercialized western shore of Cold Spring Harbor probably gained the nickname "Bungtown"—from the word "bung," the wooden plug used to stopper a barrel. *(7,8)*

(7) Bungtown Barrel Factory circa 1900.

Housing an adjunct activity essential to the flour milling and whaling enterprises of the Joneses, this two-story building stood on its own wharf on the west side of Cold Spring Harbor. Here the resident cooper and his assistants produced vast numbers of barrel staves together with bungs (tapered wooden plugs) to stopper the finished products—hence the name Bungtown for Cold Spring Harbor's busy western shore. Barrels to be used for transporting oil and bone on returning whaling voyages were tied in bundles for storage on board ship until needed.

Cold Spring Harbor L.I.

(8) Bungtown Barrel Factory circa 1885.

In this panoramic scene of Cold Spring Harbor painted in the 1880s by Edward Lange the grist mill is visible on the eastern shore (*right*) and the red and white painted Barrel Factory can be seen on the western shore (*left*).

The Jones brothers found the Cold Spring Whaling Company

Barrel manufacture at Cold Spring reached its peak after the entrepreneurial skills of the Jones brothers found ultimate fulfillment in the 1836 incorporation of the Cold Spring Whaling Company. (This was just one year after the founding of St. John's Church, the Jones family church. The church building erected a year later directly at the head of the harbor was mostly financed by the same incorporators as the Whaling Company—brothers, sisters, and cousins of John H. and Walter R. Jones plus their closest business associates.) In the midst of an ensuing national recession the whaling industry shored up the Cold Spring economy by providing a market for locally manufactured goods and for its maritime services. Between 1836 and 1858 nine whaling vessels sailing out of Cold Spring Harbor completed 38 voyages under the command of able captains commissioned by the Joneses. Thus by the middle of the nineteenth century the former haunts of the Wawepex band were barely recognizable, overrun as they now were by the textile factory foremen and their mill hands, the resident cooper and his helpers, and the shipyard supervisor and his carpenters, all busily engaged in supplying the warm woolen clothes and blankets for the long whaling voyages out, the barrels for storing the precious whale oils and bone on the return journeys, and the necessary repairs for the ships at anchor. *(9,10,11)*

(9) St. John's Church circa 1880.

Members of the Jones family of Cold Spring, together with business partners and friends, built St. John's Church in 1836 on the shore of the First Lake at the head of the harbor.

(10) Map of "Jones Industries" buildings circa 1889.

This map prepared by the Jones family genealogist John Henry Jones shows that practically all of the buildings and lands at the head of Cold Spring Harbor were under the ownership of the Jones family. Note the "Store" and "Grist Mill" on the east side of the harbor, "St. Johns, Epis Church" just to the west near the head of the harbor, and the "Woolen Factory" and "J.D. Jones Res" (now Davenport House) at the entrance to present-day Bungtown Road.

(11) Nineteenth century artifacts found along Bungtown Road.

During the first half of the nineteenth century the western shore of Cold Spring Harbor was home to many of the factory workers employed in the "Jones Industries" and their families. In many cases the workers' houses were in fact "factories" themselves, as revealed by the large-scale purchases entered in the 1850 account ledger (*center*) found in one of these homes. Resting on the left-hand page of the ledger is the sawed-off remains of a bung. After stoppering, the bung was trimmed flush with the side of the barrel. All of the items depicted here were found by Guy Cozza on the premises of Cold Spring Harbor Laboratory in or near buildings predating its founding in 1890.

Bungtown declines after petroleum flows from
Pennsylvania oil fields

Early in the second half of the nineteenth century the fortunes of the Cold Spring Whaling Company were adversely affected by the death of Walter R. Jones in 1855 and shortly thereafter by that of his brother John H. Jones in 1859. In addition in the late 1850s two Whaling Company ships were lost in the Arctic in rapid succession. The final death knell of the whaling industry at Cold Spring and other seaboard villages of the Northeast was the discovery of petroleum in Pennsylvania in 1859. Lamps burning the cleaner kerosene fuel quickly replaced whale oil lamps throughout the land. Bungtown, the former whaling era boomtown, soon looked like a deserted village. Structures that had been built to house factory workers became vacant, and everywhere along Bungtown Road factories and warehouses went unused. However, as the nineteenth century drew to a close a remarkable series of events occurred at Cold Spring to stem the tide of neglect and decay that threatened to engulf the lands and buildings on the western shore of the harbor. *(12)*

(12) Derelict buildings at Cold Spring Harbor circa 1880.

Warehouses originally built for the "Jones Industries" stood abandoned and empty in the wake of the discovery of petroleum in Pennsylvania in 1859. This precipitated the end of the whaling industry which had been the mainstay of Cold Spring's economy. "When this business became no longer profitable," commented Walter R.T. Jones years later in 1904, "the place soon appeared like a deserted village—houses became vacant, buildings unused, and everywhere neglect and decay."

CHAPTER ONE

The Brooklyn Institute of Arts and Sciences sponsors a biological laboratory at Cold Spring Harbor

1890~1924

The years after the Civil War were a boom era in the history of the United States. In this "Gilded Age" ever larger segments of the population had increased leisure opportunities and the means to enjoy them. It did not take long for the prosperous citizens of New York City to discover the natural and scenic beauties of nearby Long Island, which extended eastward for over 100 miles. Scores of affluent vacationers sailed their private yachts into secluded coves and inlets along Long Island's north shore and later purchased waterfront estates here. Other vacationers patronized a growing number of recently opened resort hotels and inns. Among the visitors to Cold Spring Harbor were those who recognized the potential of the area for conservation projects and scientific research. By the end of the nineteenth century this popular vacation spot had acquired a Fish Hatchery and a Biological Laboratory and the whaling era buildings along Bungtown Road had gained a new lease on life.

Following in Darwin's Footsteps

Biology could never be the same after the publication on November 24, 1859, of Charles Darwin's *On the Origins of Species or The Preservation of Favored Races in the Struggle for Life*. Until then the most educated people took for granted that each form of life had been created de novo by some almighty body, possibly even as recorded in the Bible. Instead, Darwin postulated that all existing forms of life arose through an evolutionary process which started from a single life form that came into existence soon after the earth itself was formed several billion years ago. Underlying evolution as postulated by Darwin were two essential processes. The first was the generation of an inexhaustible supply of genetic diversity. Darwin admitted that he did not understand how such variation was produced. The second process was the differential survival and reproduction (selection of the fittest) among the oversupply of individuals produced in every generation. Those progeny most genetically fit would give rise to the progeny of the next generation.

Darwin's personal evolution from a believer in the immutability of species into a proponent of their constant, gradual change largely arose from his observations as a naturalist on the H.M.S. Beagle. This ship of the Royal Navy spent five years (1831–1836) traversing the islands and coasts of South America and Australia. While on this voyage Darwin observed that the precise characteristics of a given species changed from one part of its geographical range to another. In particular he was fascinated by the divergences among the members of a species that occurred when they became isolated on oceanic islands like those of the Galápagos. He was equally influenced by the progressive changes in the forms of fossil plants and animals that existed in rocks emanating from different geological eras in the history of the earth. Already by the early 1840s he had committed to writing his then very heretical evolutionary theories, but they were to remain unpublished until he learned in the mid-1850s that a second British naturalist, Alfred Russel Wallace, had come to the same conclusion and was about to publish his evidence.

Within a decade of its proclamation Darwin's theory focused a virtual army of zoologists, botanists, anatomists, and embryologists on determining evolutionary relationships and inferred common ancestors. Rapidly accepted was the German biologist Haeckel's 1866 theory of recapitulation, "Ontogeny recapitulates phylogeny," which postulated that during the embryological development of an organism it passes through the morphological stages of its evolutionary ancestors. This theory enormously stimulated comparative embryological studies, which were most easily done on marine organisms whose development occurs outside the maternal body and thus is easily observable. A rush to the water to collect and observe marine organisms such as sea urchins and starfish led to the creation of a number of biological stations, mostly on saltwater oceans and seas, although many laboratories on freshwater bodies also came into existence.

The first marine biological station was established in 1872 in Naples, Italy, largely to provide space for German scientists who financed its construction and subsequent operation. Embryology was then dominated by German scientists, and they wanted to gain access to Mediterranean fauna which at that time was thought to be much richer than the fauna found in the North and Baltic Seas.

Greatly assisting the science done in the early marine stations, as well as in the home universities of their visiting scientists, was the development of new microscopes and dyes that allowed chromosomes to be seen clearly for the first time. Germ cells (eggs and sperm) with half the chromosomes of ordinary cells were discovered, and the process of fertilization was worked out. Initially, however, the role of chromosomes was an enigma, with many leading scientists believing that cells contained only one type of chromosome present in multiple copies. As the nineteenth century drew to a close, biology remained dominated by the search for further evidence in favor of the theory of evolution. The cellular mechanisms underlying it, in particular the nature of the genetic variability that fueled evolution, were to remain for the next century to handle.

Resort accommodations at Cold Spring Harbor are
plentiful

The day-trippers and vacationers who arrived at Cold Spring Harbor by packet boat in the last quarter of the nineteenth century found food and lodging readily available at several resort hotels erected around the harbor. On the western shore of the harbor was the Laurelton, a five-story hotel, complete with its own dock, that was opened by Dr. Oliver L. Jones in 1875. (Thirty years later the building was purchased by Louis Comfort Tiffany who razed it and erected his own home on the same site. Called Laurelton Hall, the Tiffany mansion was a showcase for his prodigious artistic talents in glass and mosaics until it was destroyed by fire in 1957.) Two other large resort hotels, both owned by the Gerard family, were situated on the east side of Cold Spring Harbor. The smaller, older Forest Lawn resort lay to the south of the newer, grander Glenada complex which included a bowling alley and a two-story casino. (The casino now serves as the clubhouse for the Cold Spring Harbor Beach Club.) There were several small inns as well, and rooms for paying guests were also available in private homes in the village. *(1)*

The fish culture movement gains momentum in
New York

As increasing numbers of Americans began to enjoy outdoor recreational activities wildlife conservation became an issue of the day. Typical of this trend was the pisciculture (fish culture) movement, which was dedicated to restoring rapidly disappearing game fish species to their native waters. Fish Commissions were formed in several states, including New York, in the late 1860s and early 1870s. The first hatchery of the New York Fisheries Commission was opened in 1870 in Caledonia, in the central part of western New York. Later a second hatchery was attempted at Roslyn, Long Island, but this proved unsuccessful.

Cold Spring Harbor acquires a fish hatchery

It was only a matter of time before Cold Spring Harbor's natural endowments were discovered by vacationers with a distinctly scientific bent. The pressure for a fish hatchery on Long Island undoubtedly came from wealthy Brooklyn fish dealer Eugene Blackford (1839–1905) who sat on the board of the New York Fisheries Commission and probably had vacationed at Cold Spring Harbor. Blackford owned 22 retail stands on Fulton Street and was the founder of Blackford and Company, wholesale fish sellers. He operated freezing stations in Canada, imported frozen fish from the British Isles, and is credited with introducing New

(1) Cold Spring Harbor circa 1885.

This montage by commercial artist Edward Lange highlights some of the area's resort attractions. (*Top, left to right*) "The Docks" of Cold Spring Harbor on the east side of the inner harbor; "The Laurelton," a resort hotel on the western shore of the outer harbor (later the site of Louis Comfort Tiffany's Laurelton Hall); and "The Episcopal Church" (St. John's) at the head of the harbor. (*Bottom*) Two additional resort hotels, "The Glenada" and the "Forest Lawn," both on the eastern shore (near the present-day site of the Cold Spring Harbor Beach Club), together with "The Old Mill" (Hewlett-Jones grist mill) on that same shore but close to the head of the harbor. The mill can also be seen in the panorama in the center of the montage. It is situated at the water's edge on the right-hand side. Directly to the right of the mill is the home of the miller William White. On the opposite shore in this center panorama are (*left to right*) the Fish Hatchery director's residence (now known as Davenport House) then occupied by Frederic Mather and the two small buildings being used at that time as the headquarters of the Hatchery which had been founded in 1881. The same Hatchery buildings are shown in illustration *2* on the facing page.

Yorkers to the excellence of pompano, red snapper, and whitebait as table fish. In fact, the red snapper (*Lutjanus blackfordi*) is named after him.

Attracted by the copious supplies of both fresh and salt water available at Cold Spring Harbor, the New York Fisheries Commission proceeded in 1881 to open the state's second permanent hatching station just to the west of the head of the harbor. Members of the Jones family immediately put two abandoned Woolen Factory structures at the Commission's disposal. The success of the new Fish Hatchery was virtually assured when in 1883 the noted pisciculturist, naturalist, and author Frederic Mather (1833–1900) was appointed superintendent. *(2)*

Aged 50, Mather was at the height of his career when he came to Cold Spring Harbor. He was the inventor of the Mather Hatching Cone and had supervised the United States exhibit at the International Fisheries Exhibition held in Berlin in 1880. While on a fishing expedition in the Black Forest in Germany after the exhibition he "discovered" the brown trout, and it was not long after his arrival at the Fish Hatchery in Cold Spring Harbor that he received the first large shipment of brown trout eggs exported from Germany. Although in upstate New York fish hatching focused on shad, most of the fry subsequently hatched at Cold Spring Harbor were brown trout descended from these early piscine emigres from

(2) Original buildings of the Cold Spring Harbor Fish Hatchery.

The hatchery station founded in 1881 operated out of two former Jones Woolen Factory buildings until 1887. The Fish Hatchery director's residence (now Davenport House) can be seen in outline on the extreme left.

(3) Fish Hatchery headquarters erected in 1887 by the New York Fisheries Commission.

The classes of the Biological Laboratory (founded in 1890) met at the Fish Hatchery until its own purpose-built laboratory structure (Jones Laboratory) was completed in 1893. The structure pictured here was replaced in 1958 by the New York State Conservation Department with the current yellow-brick building owned today by the Cold Spring Harbor Fish Hatchery and Aquarium, a private non-profit organization.

Germany. Mather's work was so successful that within four years the fish hatching operations were moved out of the former Woolen Factory buildings into a large, new, two-and-a-half-story wooden structure erected in 1887 just south of St. John's Church. *(3)*

A biological laboratory will come to Cold Spring Harbor

The establishment of the Fish Hatchery at Cold Spring Harbor was to have unexpected consequences for this quiet village. Fish Commissioner Eugene Blackford also sat on the board of the Brooklyn Institute of Arts and Sciences and was well aware that fellow board member Adelphi University zoology professor Franklin Hooper (1851–1914) was eager to have the Brooklyn Institute set up a seaside zoological station. Talk of Darwin's theory of evolution was in the air, and it was the latest vogue in zoological circles to study nature at its source—the sea. In his younger years Hooper had participated in the first seaside summer school for zoology in North America, the Anderson School of Natural History founded by Harvard Professor Louis Agassiz in 1873 on the rocky shores of Penikese Island in Buzzards Bay off Cape Cod, Massachusetts. Agassiz's educational philosophy, summed up in the motto "Study Nature, Not Books," captured the imaginations of nearly all the biologists of Franklin Hooper's generation. In fact, the first permanent seaside zoological station in North America, the Marine Biological Laboratory established at Woods Hole, Massachusetts, in 1888, was modeled after the Anderson School. *(4)*

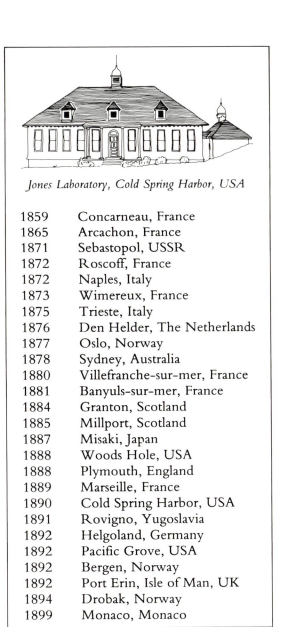

Jones Laboratory, Cold Spring Harbor, USA

1859	Concarneau, France
1865	Arcachon, France
1871	Sebastopol, USSR
1872	Roscoff, France
1872	Naples, Italy
1873	Wimereux, France
1875	Trieste, Italy
1876	Den Helder, The Netherlands
1877	Oslo, Norway
1878	Sydney, Australia
1880	Villefranche-sur-mer, France
1881	Banyuls-sur-mer, France
1884	Granton, Scotland
1885	Millport, Scotland
1887	Misaki, Japan
1888	Woods Hole, USA
1888	Plymouth, England
1889	Marseille, France
1890	Cold Spring Harbor, USA
1891	Rovigno, Yugoslavia
1892	Helgoland, Germany
1892	Pacific Grove, USA
1892	Bergen, Norway
1892	Port Erin, Isle of Man, UK
1894	Drobak, Norway
1899	Monaco, Monaco

(4) Seaside zoological stations established before 1900.

In attempting to prove Darwin's theory of evolution (first expounded in his *On the Origins of Species* in 1859) late nineteenth century zoologists went down to the sea. There they could not only study nature at its source, but also take advantage of the ease of using marine creatures as experimental organisms. Thus the first zoological stations sprang up in established fishing ports along the shores of Europe and North America.

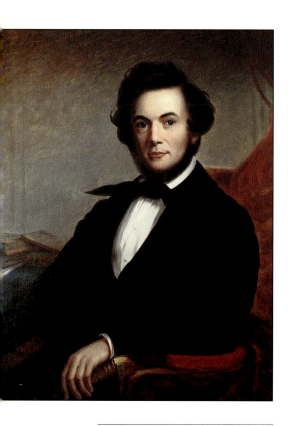

With Professor Hooper pushing for a seaside biological laboratory on Long Island it was not surprising that Eugene Blackford invited him out to Cold Spring Harbor to inspect the new Fish Hatchery there and to familiarize himself with the rich ecological diversity of the area. A luncheon meeting took place in the fall of 1889 in Director Mather's large house adjacent to the Fish Hatchery. Completed just five years earlier, this handsome residence (which today marks the entrance to Cold Spring Harbor Laboratory) was a belated replacement for the family homestead of John H. Jones which had been destroyed by fire in 1861. At the time of the fire the house was owned by Jones' eldest son John Divine Jones (1814–1895) and was undergoing repair. Although he subsequently moved to the south shore of Long Island, John D. Jones remained one of Cold Spring Harbor's largest landholders and continued to take an active interest in the affairs of his hometown, particularly through the agency of his youngest brother Walter R.T. Jones (1830–1906). Undoubtedly Walter Jones was present at the meeting in the Fish Hatchery director's house and offered to put shore-front property owned by his brother at the disposal of the Brooklyn Institute. This was probably all that was finally needed for the Institute to make the fateful decision to establish a biological laboratory right next door to the Fish Hatchery.

(5) Portrait of John Divine Jones (1814–1895) by Shepard Alonzo Mount circa 1855.

At about the time this portrait was painted John D. Jones became president of the Atlantic Mutual Insurance Company, a post he held for nearly forty years. He helped found the Biological Laboratory in 1890 and it leased its three-acre campus from him.

John D. Jones sponsors scientific research at Cold Spring Harbor

At the time of the discussions with the Brooklyn Institute John D. Jones was the chief executive officer of America's premier marine insurance company, the Atlantic Mutual Insurance Company, a position he still held at the time of his death in 1895 at age 81. He had assumed the presidency of the company when he was 42, upon the death in 1855 of his uncle Walter R. Jones, the company's founder. After almost forty years of distinguished service in this position, John D. Jones was so highly regarded in the New York financial world that on the morning of his funeral (held at Trinity Church) all of the flags on Wall Street were flown at half-mast. The highlight of his career is said to have been a meeting with President Abraham Lincoln in which, speaking for the Atlantic Company, Jones agreed to provide the necessary insurance protection for Federal shipping during the Civil War. *(5)*

When the Fish Hatchery was established at Cold Spring Harbor in 1881 the land and building that it leased for the nominal rent of one dollar per year were chiefly owned by John D. Jones. It would be likewise with the new Biological Laboratory which began operations under the auspices of the Brooklyn Institute in the summer of 1890. Eventually both of these leasing arrangements were formalized with the Wawepex Society, a family corporation founded by Jones in 1892 and named after the original Indian settlement at Cold Spring. The stated purposes of this Society were twofold: the holding of real estate and the investing of funds for scientific research.

The Bio Lab opens on July 7, 1890

Like the Wawepex Society, the Brooklyn Institute's Biological Laboratory at Cold Spring Harbor (often referred to as the Bio Lab) was also founded with two purposes in mind: to furnish a place for general biological instruction and to offer advanced students the opportunity for investigation. On July 7, 1890, the first class of students convened in the new hatchery structure of the New York Fisheries Commission since the Bio Lab did not yet have its own laboratory space. Only one course, a General Course in Biology, was offered during the inaugural season, but within ten years the number of course offerings would grow to eight. *(6)*

(6) First class of the Biological Laboratory in 1890.

The students are returning from a field trip aboard the "Rotifer."

Up-and-coming young scientists give the Bio Lab its early direction

The director during the first season at the Bio Lab was Bashford Dean, who had earned his doctorate in zoology that same year from Columbia University specializing in the embryology and paleontology of fishes. He had previously spent time at Cold Spring Harbor between 1886 and 1888 as an assistant at the Fish Hatchery. After serving as director of the Bio Lab for one summer, Dean went on to an instructorship and later a full professorship at his alma mater Columbia University. In 1903 he was made Curator of Ichthyology and Herpetology at the American Museum of Natural History and at the same time Curator of Arms and Armor at the Metropolitan Museum of Art. The first Bio Lab director was truly a Renaissance man, turn-of-the-century style.

Dean was succeeded as director in 1891 by Herbert S. Conn, a professor of bacteriology at Wesleyan University who had been on the teaching staff at Cold Spring Harbor during the inaugural season. Accompanied by his young family, Conn was in residence at the Bio Lab as its director for seven summers, through 1897.

The Bio Lab gains its own facilities

During his first summer as director Conn instituted a second course, Bacteriology, but further expansion was unlikely since the borrowed space in the Fish Hatchery building was becoming crowded. It did not take long for the Wawepex Society to recognize that the Bio Lab needed separate facilities.

After toying with the idea of financing improvements to one or more of the large warehouse structures on the grounds of the Bio Lab, the Society opted instead to commis-

sion New York City architect Lindsey Watson to draw up plans for a new purpose-built building to be used as a schoolhouse and laboratory for the biology students and their instructors. John D. Jones, who early on had taken a great interest in this project and had committed $5000 to it, was especially concerned that the new building be carefully sited so that nearby natural springs might provide all the necessary fresh water. The site chosen with this requirement in mind was right at the water's edge, adjacent to a section of shoreline that had been bulkheaded with a seawall of brownstone rubble and equipped with a dock. The entire project involved demolishing several old warehouses, including a particularly large one that stood on the site chosen for the new laboratory. A smaller warehouse to the south (later called Wawepex Building) was retained and renovated for use as a lecture hall. *(7,8)*

(7) Portrait of John D. Jones by Daniel Huntington dated 1892.

Aged 78, Jones sat for this portrait two years after the Biological Laboratory was founded and one year before he made the gift of Jones Laboratory.

(8) Jones Laboratory today as seen from the porch of Hooper House.

Jones Laboratory is the oldest seaside laboratory in North America still used for science. The wing on the right was added in 1975.

JONES LABORATORY

Lindsey Watson
1893

The Bio Lab's new facility, later called Jones Laboratory in honor of its donor, was a long, rectangular, single-story wooden structure. It was designed with a tall hipped roof for summertime coolness and topped with an elegant cupola. These roof details, together with the building's overall symmetry, classical columned entrance portico, and shingled exterior, made Jones Laboratory an early Long Island example of the Colonial Revival style of architecture that became popular after the turn of the century for churches, homes, libraries, and schools. *(9,10,11,12)*

The interior of the building, including the coved ceiling surfaces, was entirely sheathed in beaded boards. The original floor plans called for three wooden cubicles along each side of the building at the west end; a larger Director's Room and a Bacteriological Laboratory near the center of the space on the north and south sides, respectively; and a large, open, communal teaching laboratory with numerous worktables positioned between the tall windows at the east end of the building. *(13,14)*

(9) Jones Laboratory circa 1895.

Erected in 1893, the Bio Lab's first laboratory building featured many details of the Colonial Revival style—steep hipped roof, cupola (with a copper roof), shingled exterior, diamond-pane dormer windows, and a classical porch with balustrade (later lost in a storm).

(10) Jones Laboratory and dock circa 1895.

Visible on the far left is the roof of a whaling era multiple-family residence then used as a dormitory (now Hooper House). On the right is an old warehouse (later demolished), and on the far right in the distance is the former Barrel Factory and its wharf.

(11) Jones Laboratory today viewed from the south.

Note the massive brownstone chimney and to its left the low, small spring house that originally supplied the laboratory with fresh running water.

(12) Spring next to Jones Laboratory.

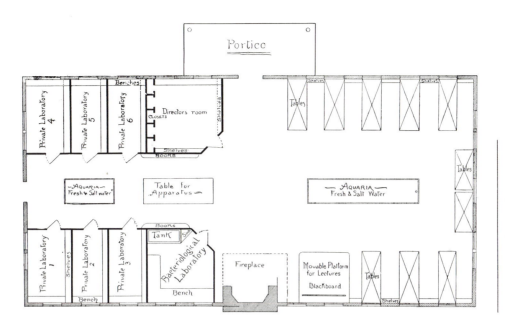

(13) Floor plan of Jones Laboratory dated 1893.

The interior of the laboratory was designed with cubicles for individual researchers on the west side of the building and communal teaching facilities on the east. (From the *Brooklyn Institute of Arts and Sciences Yearbook* for 1893–1894)

(14) Interior of Jones Laboratory circa 1895.

In this view looking toward the west (side) entrance the Bacteriological Laboratory is on the left and the Director's Room is on the right.

(15) Wawepex Building circa 1900.

Originally built as a warehouse for the "Jones Industries" at the end of the first quarter of the nineteenth century, it was adapted for use as a lecture hall for the students at the Bio Lab in the last quarter of the century. In this view Wawepex Building is partially obscured by the roof of Jones Laboratory in the foreground.

WAWEPEX BUILDING
(Built circa 1825)
Leased
1893

The abandoned whaling era warehouse adjacent to Jones Laboratory was also put at the Bio Lab's disposal as part of its leasehold agreement with the Wawepex Society. This two-and-a-half-story shingled building was easily adapted to provide a darkroom and lecture hall on its bottom floor; at this level the building was mostly below grade and had windows only on the side facing the harbor. The scientists thus put Wawepex Building to a good new use to supplement the facilities available in Jones Laboratory. *(15,16,17)*

(16) Wawepex Building in the 1920s.

Laboratories had been installed inside by this time. The roof of the Animal House (built in 1914; now McClintock Laboratory) is visible on the right in the background.

(17) Wawepex Building today.

This former warehouse was once again adaptively reused in 1971 when it became a dormitory.

*The number of courses grows thanks to Jones
Laboratory, and ecological studies begin*

The completion of Jones Laboratory made it possible for additional courses to be offered, and their number and scope grew each year. In the summer of 1900, ten years after the Bio Lab was founded, the following eight courses were available: High School Zoology, Comparative Anatomy, Invertebrate Embryology, Variation, Cryptogamic Botany, Phanerogamic Botany, Bacteriology, and Microscopic Methods. In 1901 a course on Ecology was taught for the first time. This was one of the earliest courses of this type offered anywhere, and it ushered in an era of ecological studies of the Cold Spring Harbor environs. A landmark in the history of these studies was the publication in 1915 of Duncan S. Johnson and Harlan B. York's *The Relation of Man to Tide Levels: A Study of the Factors Affecting the Distribution of Marine Plants*. Based on research that spanned many summers at the Bio Lab and included numerous off-season study visits, this monograph focused on the environmental factors that affected the distribution of the various marine plants that grew in the intertidal zones of the inner harbor at Cold Spring and the Sand Spit that forms its northern boundary.

While completing their study of the intertidal plant life of Cold Spring Harbor Johnson and York were the instructors in the Cryptogamic Botany course, which dealt with marsh grasses and other non-flowering plants (as distinct from phanerogamic, or flowering, plants). Another researcher involved in the summer teaching program at the Bio Lab was Henry S. Platt. A longtime instructor in the Comparative Anatomy course, Platt published his *Manual of the Common Invertebrate Animals* in 1919. In the early days of the Bio Lab, however, much of the teaching was done by the director himself. There were only one or two dozen students, and the staff was correspondingly small.

Charles Davenport takes charge of the Bio Lab

Harvard-trained zoologist Charles Benedict Davenport (1866–1944) succeeded Herbert Conn as the third director of the Bio Lab. Davenport was working as an instructor at Harvard when he was tapped for the summer job on Long Island in 1898. Like Conn, he found that the Bio Lab directorship involved less administrative work and more teaching responsibilities. In the latter activity he would be aided by his wife Gertrude Crotty Davenport, who had been studying at Harvard for her own doctorate at the time of their marriage. Charles Davenport was to serve as director of the Bio Lab for the next twenty-five years, a period that saw great changes in the way science was done at Cold Spring Harbor. An avid experimentalist and prolific writer, particularly on the subject of Mendelian genetics, Davenport provided a charismatic presence during both the good times and the precarious ones at the Bio Lab.

Housing problems are lessened through the lease of Hooper House

Finding decent housing and board for the students and instructors was not easy in the early days, although the latter would improve with the arrival of the Davenport family. Once Jones Laboratory was completed there was ample laboratory space at the Bio Lab, but there were only a few buildings available on the grounds for housing. Many of the students in the summer courses had to look to Cold Spring Harbor village across the harbor for lodgings and at the beginning for board as well. Both of these needs were met by two local taverns, Van Ausdale's and the Village House. This far from ideal situation began to improve when the Wawepex Society decided to put various old structures on the site at the biologists' disposal, one of the first being the building now called Hooper House.

HOOPER HOUSE
(Built circa 1835)
Leased
1893

One of the first buildings on the grounds of the Bio Lab to be used for summer housing was an abandoned multiple-family dwelling that dated back to the heyday of the "Jones Industries" at Bungtown. Originally built for textile workers and their families, this large, but plain, two-and-a-half-story shingled structure with matching wings at each end began to be used as a men's dormitory in 1893. In the late 1890s part of the ground floor of the house became a dining hall for the teachers and students. It was later given the name Hooper House in honor of Franklin Hooper who had helped found the Bio Lab. *(18,19)*

(18) Hooper House circa 1900.

Built in the 1830s as a multiple-family dwelling for textile workers, in the 1890s it was used in the summer as a dormitory and dining hall for Bio Lab students.

(19) Hooper House today.

Since 1960 it has been a year–round apartment residence with a dormitory on the top level.

Fresh produce for the biologists' tables

When the Davenports came to the Bio Lab they personally took charge of the catering operation in Hooper House and would write ahead in the spring to neighboring farmers to ensure an adequate supply of produce for the biologists' tables. James Wheeler, one of the people with whom Davenport corresponded, planted potatoes at his request and also suggested additional lodging places in the area. The Wheelers themselves made rooms available in the house that they rented from the Joneses. Similar in appearance to Hooper House and also originally built for workers' housing, the Wheeler residence (later known as Williams House) was situated across Bungtown Road from Hooper House and a little to the north of it. It did not actually lie within the confines of the small parcel of land (barely three acres) that the Bio Lab leased from the Wawepex Society, although a small cottage did. For a brief while the Davenports made their home in this cottage (later called Osterhout Cottage).

OSTERHOUT COTTAGE
(Built circa 1800)
Leased
1893

The small and very old story-and-a-half shingled cottage that stood on the Wawepex land was used as a men's dormitory before becoming the Bio Lab director's house in 1898. Many years later it was named Osterhout Cottage, after W.J.V. Osterhout, a Bio Lab trustee in the 1920s and 1930s.

With the bedrooms situated in the tiny attic story these accommodations could not have been very comfortable on steamy July nights. It is therefore not surprising that Davenport and his family were thrilled to move out of this cottage and into the former residence of Fish Hatchery director Frederic Mather (eventually named Davenport House) which the Wawepex Society put at their disposal in 1903. (20,21)

(20) Osterhout Cottage circa 1880.

Built in the early nineteenth century, this small cottage was used by the Bio Lab in the 1890s first as a dormitory and then as the director's house.

(21) Osterhout Cottage
today.

The house was reconstructed
in 1969.

DAVENPORT HOUSE

(Built in 1884 as the Mather house)
Acquired
1903

The home to which the Davenport family (two daughters and later a son) moved in 1903 was built on the site of the country residence of the Laboratory's founder John D. Jones which had been destroyed by fire in 1861. Before he subsequently moved to the south shore of Long Island he ordered the necessary materials for rebuilding the house. Strangely enough, nothing more was done for twenty years, not until after the Fish Hatchery was established at Cold Spring Harbor. When Director Mather arrived in 1883 he, like all of the other scientists who later migrated here in his footsteps, needed a place to live nearby. *(22)*

The Jones family homestead was quickly rebuilt, presumably along the lines of the former dwelling. Completed in the fall of 1884, the new residence was two-and-a-half stories high and had a full basement. The interior layout was similar to that of the typical pre–Civil War Long Island farmhouse. The rooms were arranged according to the central hall plan: a dining room and parlor on the west side and a double parlor on the east side. There was also a two-story kitchen wing on the west side. *(23)*

Although the floor plan was old-fashioned, the finishing details of the res-

(22) Davenport House in 1904.

Built in 1884 for the Fish Hatchery director, it became the home of Bio Lab director Charles Davenport in 1903. The trench in the foreground is for a water connection. The small building in the background is an ice house.

idence were in up-to-the-minute late nineteenth century High Victorian style. All of the first floor reception rooms had fireplaces of dark, marbleized slate (including two in the double parlor) and they all had wooden wainscoting, stained a deep rich mahogany color, up to a height of four feet. Other late Victorian features included most of the interior trim and the design of the windows, which had borders of decorative small panes on the upper sashes. The decorative half-timber work on the exterior of the building was also typical of the day, as were the two projecting tower-like elements, centered on the front and back of the house, where this decorative detailing was most prominently featured. *(24,25)*

(23) Floor plan of the basement of Davenport House as remembered by Jane Davenport Harris de Tomasi in 1979.

The Davenports took personal charge of the room and board arrangements for the Bio Lab students. The cellar of their house apparently served as central supply for the catering enterprise.

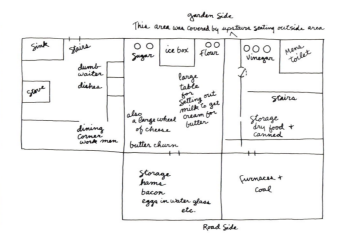

(24) Davenport House in the 1940s.

After Davenport retired in 1934 the house was used as a dormitory.

(25) Davenport House today.

The house—and the original colors—were restored in 1980.

*1903 is a banner year at the Bio Lab, reports
Director Davenport*

Moving into the grand house at the entrance to the Bio Lab was one of the highlights of 1903 for Director Davenport. In his Report to the Board of Managers of the Bio Lab for that year he singled out this move as one of the factors contributing to his sense of satisfaction with the 1903 session: "It gave a dignified residence for the director and his family in which it was possible to pay some of the social debts that had accumulated, and a social center for the students, and a suitable place in which visiting naturalists could be entertained." It also freed up the small cottage for use once again as a men's dormitory, which helped not only to alleviate the chronic housing shortage, but also to generate rental income for the Bio Lab. Problems about the future administration of the Bio Lab, which had been brewing for several years, were also successfully resolved in 1903.

*Columbia University will NOT take over the
Bio Lab!*

Overcrowding and lack of funds were ever-present concerns at the Bio Lab in its early days, but these problems came to a head in the early Davenport years. Not long after the tenth anniversary of the Laboratory's founding there was a growing concern about the number of students that could reasonably be accommodated each summer on the small campus that the Wawepex Society leased to the Brooklyn Institute on a yearly basis, and members of the Society even discussed among themselves the possibility of not renewing the lease of the grounds and buildings to the Institute.

The thought was that perhaps Columbia University, alma mater of most of the Joneses and the source of many summer students and professors, might manage the Bio Lab better than the Brooklyn Institute. Bio Lab director Charles Davenport and his Board of Managers were firmly committed to the Brooklyn Institute, and it to them, especially in the person of Franklin Hooper, a general director of the Institute and also the Bio Lab's secretary. In a letter dated July 29, 1902, Townsend Jones, the Wawepex Society member who lived closest to the Bio Lab, informed his uncle Walter R.T. Jones, Governor of the Wawepex Society, that he was "inclined to believe that the Brooklyn Institute intend having a laboratory at Cold Spring whether we help them or not by giving a renewal lease. It looks to me as if they want to stick to us if we let them."

The Joneses were probably impressed by the barrage of impassioned letters from Charles Davenport outlining the many reasons why the Wawepex Society should renew the lease of the Bio Lab land to the Brooklyn Institute rather than lease it to Columbia University. In a letter to W.R.T. Jones dated July 13, 1902, Davenport stressed the desirability of the Bio Lab's remaining in a neutral position and "not subject to University rivalries." He pointed out that when the University of Chicago offered to take charge of the Marine Biological Laboratory at Woods Hole the overwhelming opposition that eventually defeated the proposal was based on a fear that the Woods Hole Laboratory would no longer be able to count on the support of other universities. The six scholarships that had already been established at Cold Spring Harbor by outside institutions would certainly be withdrawn if a single university, such as Columbia, were to take over the administration of the Bio Lab.

Davenport also noted the excellent past management of the Bio Lab by the Brooklyn Institute and cited the wonderful reputation Cold Spring Harbor had achieved as a place to do science—"in a book published in Germany giving a list of scientific localities of the world, Cold Spring Harbor is placed in alphabetical position in the midst of Berlin, Boston, Brussels, Edinburgh, London, New York, Paris and Vienna." This he maintained was the principal reason for the record number of biologists resident at the Bio Lab in the summer of 1902—there were 42 students that year and 16 instructors/researchers.

The adoption of certain rules and regulations, mostly regarding the student body, formed the basis of the understanding that was soon reached between the Wawepex Society and the management of the Bio Lab. As noted in Davenport's letters, the teaching period was to be limited to July and August, and henceforth the students were to be at least 21 years of age, have a college degree or the equivalent, and be capable of pursuing research. In addition, the Board of Managers was to raise an endowment of at least $15,000. Finally, the board was charged with the most important responsibility of all, "to secure the appropriation from the Carnegie Institute" which Davenport was already vigorously pursuing.

Within a year Davenport won the sought-after commitment from the Carnegie Institution to establish a year-round Station for Experimental Evolution directly next door to the Bio Lab. This was to have an enormously stabilizing effect on the relationship between the Wawepex Society and the Bio Lab. In 1903 the Society extended the term of its lease to the Brooklyn Institute to ninety-nine years to match that of the adjacent eight-acre parcel newly leased to the Carnegie Institution of Washington.

Charles Davenport was the obvious choice to head up the new sister institution to the Bio Lab, and as part of the lease agreement made between the Wawepex Society and the Carnegie Institution, Davenport and his family were entitled to take up residence in the former house of the Fish Hatchery director. Davenport also continued on as Bio Lab director, and his residing in such a gracious home lent much cachet to the summertime activities of the Bio Lab.

Neighbors help the Bio Lab make ends meet

Unlike the Carnegie Station (formally opened in 1904; see Chapter Two) the Bio Lab did not have a large annual budget, usually under $3000, of which about half went for salaries for instructors. The number of instructors each summer ranged from seven or eight to a dozen or so and their salaries were low, but one of the privileges of these positions was the opportunity to do independent research. In addition to salaries, other Bio Lab expenses were, in decreasing order of magnitude, publications, maintenance, repairs to the launch and the collecting boat, and chemicals and apparatus. Certain types of laboratory equipment could be borrowed on short-term loan from nearby universities; Adelphi University, for example, loaned microscopes. Also helpful in keeping to the budget were the subsidies received from outside organizations such as the American Association for the Advancement of Science, which underwrote two research positions starting in 1895. Tuition for the season varied from $20 to $25, depending somewhat on the yearly funding from outside sources, mainly private contributions called "subscriptions."

Contributors in the early days included members of neighboring families, such as the Joneses, the de Forests, and the Tiffanys, and out-of-town subscribers, including the Honorable Seth Low, president of Columbia University and two-time mayor of New York City. In 1904 Gertrude Davenport took out a subscription for $130, which was $30 more than the contribution that year from wealthy neighbor Walter Jennings, a founder of Standard Oil. Mrs. Davenport, a trained microbiologist, also taught a combined course in Embryology and Microscopical Techniques for many years, in addition to supervising the room and board arrangements for all the Bio Lab's summer residents.

Marine biology flourishes at Cold Spring Harbor at the turn of the century

A not inconsiderable expense for any seaside laboratory was the maintenance of a boat for collecting marine organisms for use as laboratory specimens. The Bio Lab's first vessel (shown earlier in illustration 6) was a naphthalene-powered launch called the "Rotifer" (appropriately named after a class of minute aquatic invertebrate animals that have rotatory organs used in swimming). During Davenport's second summer at Cold Spring Harbor the Bio Lab was privileged to enjoy the use, for a day, of the "Fish Hawk," a trawler on loan from the U.S. Fish Commission Laboratory at Woods Hole. "Five hauls of the dredge were made...[and] the sterility of the centre of the Sound as compared with the rocky points was made evident," wrote Davenport in his 1899 Report, and "two weeks later, through the kindness of Oliver L. Jones, the school enjoyed the use of the steamer 'Grit' to visit the Connecticut shore," where several more successful dredges were made.

After being at the Bio Lab for five years Davenport wrote in his 1903 Report that it was time "that the Laboratory acquire a boat built on the plan of an oyster boat and especially fitted for dredging. The 'Rotifer' is fast going to pieces and our dredging trips are restricted each year to nearer and more protected localities."

Gertrude Stein, medical student, is enrolled at the Bio Lab

The roster of "Students Engaged in Research" at the Bio Lab in 1899 as listed in the *Brooklyn Institute of Arts and Sciences Yearbook* included a familiar name, although not in a biological context—"Gertrude Stein, A.B. (Radcliffe College), Student, Johns Hopkins University." In fact, there is correspondence in the files at Cold Spring Harbor Laboratory between Leo D. Stein and Charles Davenport regarding the 1899 summer research plans of Stein, a Ph.D. in biology, and his sister Gertrude, a medical student. In a letter dated March 11, 1899, that ends with the words "My sister sends her regards to Mrs. Davenport & yourself to which I add mine," Stein inquired about the availability at Cold Spring Harbor of "nemertean material" (vividly colored, burrowing marine worms) for himself and a microtome (an instrument for finely slicing specimens for viewing under a microscope) for his sister. Not long afterward the Steins, sharing a keen interest in aesthetic matters, moved from Baltimore to Paris, their passion for art presumably surpassing their love for biology.

The importance of a healthy student body

Unfortunately Gertrude Stein's later writings did not include details of the scientific life Cold Spring Harbor style, but some of the flavor of the summer life of a biologist at the turn of the century can be gleaned from the correspondence files at Cold Spring Harbor Laboratory. Director Davenport took a great interest in the physical as well as intellectual health of the participants at the Bio Lab's courses. This is clearly evident in the correspondence he maintained for some thirty years with Henry W. de Forest, the Bio Lab's neighbor to the north, regarding swimming privileges at the Sand Spit. Although chiefly owned by the de Forests, the Sand Spit also partly ($1/16$) belonged to the Bio Lab through its Wawepex Society connections, and thus Davenport felt that the biologists were entitled to swim there. *(26)*

The complaints about the behavior of the scientists that de Forest conveyed to Director Davenport over the years included too much noise as they made their way across his land to the beach on the Sand Spit just below his house, too many swimmers at the beach, and swimmers sunning at the wrong end of the Sand Spit beach. To keep down the noise and numbers, de Forest proposed in 1914 that the swimmers row to the Sand Spit from the dock next to Jones Laboratory rather than walk down to the end of Bungtown Road. Prompted by Davenport, he contributed $100 that same year toward the cost of purchasing five cedar rowboats for the Bio Lab.

Some of the complaints about beach activities may have reached the ears of Wawepex Society members. In July of 1902 Davenport felt compelled to convey to Walter

(26) The Sand Spit.
For over 100 years summer visitors to the Laboratory have enjoyed swimming here.

R.T. Jones some explanation regarding the activities of the Bio Lab students: "All are at work at 8 a.m. and work until 5 or 6 p.m. and often during the evening. Every one of the days of work is crowded full to the brim with duties very willingly performed. If the students go bathing on hot days at 5 o'clock it is because the bath enables them to work better. ...I assure the Society that no loafer or mere pleasure seeker is admitted to this Laboratory if I know it."

A more pressing concern, however, was the threat of disease, especially in the early 1900s. Dysentery and malaria were both problems when Davenport first came to the Bio Lab. He wrote in 1902 of the measures that were being taken to combat these problems. "Considering it probable that the spread of dysentery was due to flies breeding in the latrines and later visiting the food, the contents of the latrines on the Laboratory grounds and on those of the neighbors were well covered with kerosene oil." The result was that there were fewer flies that summer than previously and no cases of dysentery.

Unfortunately the malaria problem was not so easily solved. After noting in his 1902 Report to the Board of Managers that "malaria is now known to be spread by the *Anopheles* mosquito," Davenport reported that he had entrusted one of the staff scientists with the task of doing away with the mosquito breeding places around the Laboratory. This trusted colleague "identified many breeding places but found extensive drainage necessary to render these places innocuous; this work he could not do...." Even after "suspected spots" at the Bio Lab were treated with petroleum and the mosquito population in the immediate area appeared to be on the decline, two new cases of malaria developed.

William Matheson, mosquito enemy No. 1, becomes the Bio Lab's friend

At about this time the Bio Lab received a visit from a public-spirited neighbor, William John Matheson (1856–1930), who contacted the Laboratory on behalf of the North Shore Improvement Association. He hoped the Bio Lab could assist the Association in its campaign to exterminate mosquitoes from the north shore of Long Island. Matheson himself had purchased a large estate, Fort Hill, in the Lloyd Neck area of Huntington in 1900, although his concern about the mosquito problem may have come from his involvement with real estate development in Florida. At one time he owned the entire island of Key Biscayne.

A chemist, financier, and philanthropist, Matheson was a leader in the development of the synthetic dye industry in the United States. In 1921 he merged the company he founded, National Aniline Chemical Company, with others to create Allied Chemical & Dye Corporation. His best-known philanthropy was the establishment of the Matheson Commission for the international study of sleeping sickness. Its first report, *Epidemic Encephalitis,* was published in 1927.

Happily Director Davenport was able to report that during the 1903 season "there was no case of illness (except one slight one due to overeating at the clam-bake)," but it is possible that this was due to a lack of rain that summer. In the meantime William Matheson had joined the Bio Lab's Board of Managers, the beginning of an association that would last twenty years.

President Matheson gets Blackford Hall built

Matheson quickly rose to the presidency of the Bio Lab's Board of Managers, succeeding Eugene Blackford, one of the Laboratory's founders, upon his death in 1905. Matheson saw as one of his first responsibilities the successful completion of the building that Mrs. Blackford wished to erect on the grounds of the Bio Lab as a memorial to her husband. It was the kind of building that had been on every Bio Lab director's wish list from the beginning—a proper dormitory and dining hall for the students.

The site for the building was first proposed in a letter to Franklin Hooper dated July 11, 1905, from Frederick Law Olmsted, Jr., of Olmsted Brothers. In the late nineteenth and early twentieth centuries this noted Brookline, Massachusetts, firm of landscape architects designed more than 50 estates on Long Island (and hundreds elsewhere) as well as parks and university campuses. Olmsted suggested to Hooper that the Bio Lab negotiate with the Carnegie Institution and the Wawepex Society for a desirable building site on some of the higher ground that was on lease to the Carnegie Institution. While these negotiations were being completed, President Matheson consulted with Gardner & Howes, a New York firm specializing in concrete construction which he had retained to design several structures of reinforced concrete on the grounds of his Lloyd Neck estate. It must have been Matheson who proposed using concrete for the new building. He was familiar with this material and being a farsighted and practical man would particularly have valued both its fireproof and hygienic qualities. He obviously succeeded in getting Mrs. Blackford to share this viewpoint, and soon preparations were under way for the new dining hall and dormitory.

In addition to cement and large quantities of pebbly aggregate (which were delivered by barge to the Jones Laboratory dock) the on-site manufacture of concrete for the new building also required huge supplies of water and sand. Both were readily available on the Bio Lab grounds, the sand coming from the hill opposite the building site proposed by the Olmsted firm. In addition to the workers needed to mix the concrete, skilled carpenters were required to fabricate the wooden formworks (molds) into which the concrete mixture was poured. The grain of the wood used to make the forms left a permanent imprint on both the interior and exterior walls of the building, giving an interesting texture to the bare concrete. *(27)*

(27) Blackford Hall at the time of its completion in 1907.

The marks of the wooden formworks (molds for the concrete) are clearly visible on the finished building. Comparison of the photograph and the original floor plan in this plate published in *Concrete Country Residences* (1907) shows that the actual building had a longer "Dining Room" than was originally designed.

DORMITORY AT COLD SPRINGS HARBOR, L. I.
Solid Reinforced Concrete

Gardner & Howes, Architects

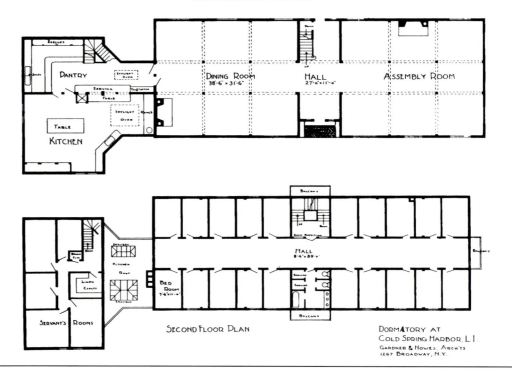

SECOND FLOOR PLAN

DORMATORY AT
COLD SPRING HARBOR, L. I.
GARDNER & HOWES, ARCH'TS
1267 BROADWAY, N.Y.

BLACKFORD HALL
Gardner & Howes
1907

The building erected at the Bio Lab in memory of Eugene Blackford was unique for Cold Spring Harbor and unusual for Long Island as well, being one of the first large residential structures built entirely of poured-on-the-site concrete reinforced with iron rods. In fact, it was completed in the same year (1907) as Frank Lloyd Wright's Unity Temple in Oak Park, Illinois, which is widely recognized as an im-portant early use of reinforced cast-in-place concrete as both an architectural and a structural material. Before this time, reinforced concrete was used mainly for foundations and for buildings of a utilitarian nature. By the end of the first third of the twentieth century, however, it was considered a desirable building material for private residences, even prestigious country homes. *(28)*

(28) Blackford Hall today.

An early example of residential construction using iron-reinforced poured-on-the-site concrete, Blackford Hall was designed as a dormitory and dining hall and it serves these same functions today.

(29) Blackford Hall circa 1910.

When viewed from the back the building exhibited a Classical purity of line—almost Grecian. The hill where the sand for the building was excavated can be seen in the background; the poultry houses of the Carnegie Station for Experimental Evolution (founded in 1904) are in the foreground.

Residential architecture was a specialty of Blackford Hall architect Robert W. Gardner (1866–1937) and he pioneered the use of reinforced concrete floors and walls for homes. An archaeologist by avocation, he was also the author of *The Parthenon: Its Science of Forms* (1925). The architectural style of Blackford Hall, if it can rightly be assigned one, is perhaps closest in spirit to the solid simplicity of the Greek beloved by Gardner. His partner Benjamin A. Howes (1876–1952) was a construction engineer and architect who worked in the mining industry before entering the housing field. *(29)*

The design of Blackford Hall, which was featured in the book *Concrete Country Residences* published by the Atlas Portland Cement Company in 1907, called for a two-story rectangular building with an entrance and stair hall located at the center on the ground floor. A commodious dining room, connected at its north end to a pantry and kitchen wing, was on the north side of the center hall. A large assembly room, with an impressive fireplace, comprised the south side of the building on the main floor. Twenty small bedrooms were arrayed on either side of the hallway on the second floor. *(30,31,32)*

(30) Assembly Room of Blackford Hall circa 1907.

The portrait is of Eugene Blackford (1839–1905). A member of both the New York State Fisheries Commission and the Brooklyn Institute of Arts and Sciences, Blackford was instrumental in having the Brooklyn Institute found its Biological Laboratory right next door to the Fish Hatchery he had previously helped found at Cold Spring Harbor. The painting still hangs in Blackford Hall today.

(31) Dining Room of Blackford Hall circa 1907.

The tables are set for a banquet, possibly celebrating the visit of the Seventh International Zoological Congress in 1907.

(32) Two meal cards dated 1908 and 1932 from the Blackford Hall Dining Club.

The price differences were due both to inflation—breakfast was 25 cents in 1908, 50 cents in 1932—and to changing dietary customs—judging from the cost, the main meal was at midday in 1908 and at the end of the day in 1932.

Leaky roof problems are solved, eventually

Although the workmanship on Blackford Hall was generally good, almost from the beginning there were problems with leaks in the roof, apparently due to minute cracks in the concrete. After eight summers of damp clothes caused by leaks in the bedrooms on the second floor and many failed attempts to patch the roof using various patented processes, a rubberized roof treatment applied by the Casmento Company in 1914 eventually did the trick.

Fortunately the leaks did not keep visitors away from Blackford Hall. In 1907 it was the scene of the Seventh International Zoological Congress, and shortly thereafter Director Davenport hosted two more large groups there: the Society for Experimental Biology and Medicine and the American Society of Naturalists. This large, new, multipurpose building also made the Bio Lab a more pleasant place for the course instructors, investigators, and students who were at Cold Spring Harbor for the entire summer. "During the last three years the whole tone of the student feeling has changed," wrote Davenport in 1910. "Our added dignity"—he cited Blackford Hall as the cause—"commands their increased respect and loyalty; and the word is passed around that a summer at Cold Spring Harbor affords not only the best of biological instruction, but an environment of beauty and conditions of living that are far more than tolerable."

Is the Bio Lab really a part of the Brooklyn Institute?

Just as the Blackford Hall roof leaks were finally being fixed a potentially more serious administrative problem arose. The Bio Lab had celebrated its 25th season under the sponsorship of the Brooklyn Institute when, as Director Davenport related in his 1914 Report to the Board of Managers, the Institute was reorganized into just three departments: the Museum, the Botanic Garden, and the Department of Education. Amazingly in this reorganization "the Biological Laboratory, which in actual operation antedates both the Museum and the Garden, is not recognized at all [as a department]. ...Indeed, some of the trustees are inclined to doubt if the Laboratory really belongs to the Institute." Just months before the Bio Lab had lost its great patron and staunch supporter on the Brooklyn Institute board when Franklin Hooper passed away. "Our present anomalous position can not be advantageously maintained," concluded Davenport, and he recommended that a delegation from the Bio Lab's Board of Managers meet immediately with representatives of the Brooklyn Institute.

In addition to being faced with this administrative crisis, the Bio Lab was also experiencing financial difficulties. Although this was not a new situation, it put the Bio Lab in a bad light vis-à-vis a normalization of relations with the Brooklyn Institute. Davenport therefore felt it was necessary for the Board of Managers to secure an endowment of $25,000 if the Bio Lab were to continue to exist and to grow. "The Laboratory, while it has made a fair growth, is challenged to greater development by the advance made in the past year by our friendly rival at Woods Hole. To stand still in all this reorganization and progress is to be thrust out of the ranks and to be left hopelessly behind"—a dire prognosis for the Bio Lab.

When the Bio Lab's Board of Managers failed to raise the necessary endowment in 1915 Davenport threatened to resign, sadly remarking that "if I have not made the Laboratory seem worthy of this much support, I have failed ignominiously and no other considerations can dim my duty of helping find my successor as director." Two years later there was finally good news to report on both the administrative and the financial fronts: the Bio Lab had become the fourth coordinate department of the Brooklyn Institute and an endowment fund of $25,000 had been successfully raised from such "friends of the Laboratory" as Mrs. E.H. Harriman, August Heckscher, Walter Jennings, William J. Matheson, Louis Comfort Tiffany, Mrs. Willard D. Straight, and Albert Strauss.

CHAPTER TWO

The Carnegie Institution of Washington establishes a Station for Experimental Evolution at Cold Spring Harbor

1904~1940

From the moment Charles Davenport arrived on Long Island in the summer of 1898 to assume the directorship of the Bio Lab he took a broad view of the future of biology at Cold Spring Harbor. His earliest reports to the Bio Lab's Board of Managers revealed that he felt that the Laboratory's long-term prospects were far from good unless a year-round program of research could be started to supplement the summer teaching program. Six years later this goal was realized when the Carnegie Institution opened its Station for Experimental Evolution at Cold Spring Harbor in 1904.

Not long after the successful culmination of his vigorous efforts to secure this program Davenport's own research efforts turned in the direction of human genetics and eugenics. His subsequent writings on eugenics established him as a leader in this field in the United States. Ultimately the eugenics studies at Cold Spring Harbor, based on flawed scientific principles, were abandoned while the more significant genetics research being conducted on the grounds of the Carnegie Station came to the fore.

Expanding the Scope of Mendel's Laws

Only six years after the publication of Darwin's *On the Origins of Species* the Austrian monk and plant breeder Gregor Mendel published experiments revealing the existence of hereditary factors that we now call genes. From his breeding of varieties of the common garden pea Mendel concluded that morphological traits like seed color and shape were controlled by pairs of hereditary determinants, one coming from the male parent and the other from the female parent. These seminal results, published in an obscure botanical journal in 1865, attracted little attention. That Mendel's rules were so easily neglected reflected the fact that his discoveries were made before the role of chromosomes in heredity was established. Unknown then were both mitosis—the ordinary cell division process in which each chromosome is duplicated just prior to cell division—and meiosis—the special cell division process that creates egg and sperm cells bearing exactly half the number of chromosomes found in ordinary cells.

By the turn of the century, however, the details of chromosome duplication and segregation were known and they provided a physical basis to interpret the new breeding experiments of the botanists Hugo de Vries, Carl Correns, and Erich Tschermak, who in 1900 independently published results confirming Mendel's results. In his classic 1902 paper the American Walter Sutton explained Mendel's results by the assumption that the genes are located on the chromosomes. In 1905 the first traits were assigned to specific chromosomes with the discovery at Columbia University by Edmund B. Wilson and his student Nettie Stevens of how sex is determined by the independent segregation of the X and Y chromosomes. Although most chromosomes are present in two morphologically identical copies, males contain only one X chromosome, with the missing X being replaced by the Y chromosome. Maleness is determined by one or more genes carried on the Y chromosome.

The rediscovery of Mendel's laws immediately gave rise to a burst of experiments extending these ideas to a variety of different species, animal as well as plant. The Englishman Gregory Bateson bred chickens at Cambridge, William E. Castle bred mice at Harvard, and Thomas H. Morgan began experiments on the tiny fruit fly *Drosophila* at Columbia. Soon experimentation with *Drosophila* began to dominate genetics, for the fruit fly was easy to breed in the laboratory, a new generation appearing every two to three weeks. Many crosses could be made quickly, and it was in the fly room at Columbia that the linear arrangement of genes along chromosomes was first demonstrated. Morgan was assisted in these experiments by three very talented students: Alfred H. Sturtevant, Calvin Bridges, and Hermann J. Muller. Each went on to play a major role in the development of genetics up to the start of World War II.

The processes by which genes change from one form to another are known as mutations. Mutant genes generally arise at very low frequencies, and a great advantage of using *Drosophila* for genetics research was the ability to screen many thousands of flies for new mutant genes. Thus almost from the moment Mendel's laws were rediscovered attempts were made to find agents that increased the frequency of gene mutations. Real success did not come until 1926 when Hermann J. Muller, then in Texas, showed that X rays are powerful mutagenic agents. Soon virtually every major genetics laboratory contained X-ray generators to produce the needed new mutant genes.

In time it became clear that the larger *Drosophila* chromosomes each contained many hundreds of genes, but whether or not mutations had been found in most of them was not known. Individual chromosomes were just resolvable using light microscopy, and there was little hope of ever seeing single genes. Thus there was great excitement in the early 1930s when the salivary glands of many insects, including *Drosophila,* were found to contain giant chromosomes approximately 1000 times thicker than those of other cells. Each salivary chromosome had banded structures, with the largest containing over 1000 distinguishable bands. Initially it was suspected that each band represented a distinct gene, and very detailed descriptions of these chromosomes were made. The most detailed of these descriptions was that done by Calvin Bridges working at Cold Spring Harbor. These beautiful maps could not by themselves tell us where the genes were as long as we did not know the molecular species out of which they were constructed, and until the end of the World War II the worlds of the chemists and the geneticists were too far apart to bridge.

Davenport pursues the Carnegie Institution

At the time of his appointment to the directorship of the Bio Lab Davenport was teaching at Harvard University, but the following year he moved to the University of Chicago as an assistant professor. Thus it was from Chicago that he waged his campaign to make Cold Spring Harbor the site of a year-round research effort in the field of evolutionary biology. Davenport concentrated his fund-raising efforts almost exclusively on the Carnegie Institution of Washington which had just been founded (1901–1902) to advance the cause of science across the United States. The Carnegie Institution had almost immediately begun seeking a site in the Northeast at which to establish a research institute for experimental biology. For many months it seemed as though this facility would be established at the Marine Biological Laboratory at Woods Hole, Massachusetts. In the end the concern of the trustees at Woods Hole that their Laboratory might lose its independence won out, and the Carnegie Institution's offer to fund increased research there was turned down. Almost overnight Charles Davenport's dream for Cold Spring Harbor was realized when the Carnegie Institution decided instead to found a Department of Experimental Evolution right next door to the Bio Lab and to make the Bio Lab director the head of their new experimental Station.

The Carnegie Institution dedicates its Station for Experimental Evolution at Cold Spring Harbor

On Saturday, June 11, 1904, a gala dedication ceremony took place on the grounds of the Bio Lab to mark the formal opening of the Carnegie Institution's Station for Experimental Evolution. Fifty guests came from New York City to join fifty neighbors at a luncheon in the director's residence. Charles Davenport, as director of the new Station, opened the program of distinguished speakers with a rousing Introduction: "We do not celebrate here the completion of a building, we are dedicating no pile of bricks and lumber—rather, this day marks the coming together for the first time of the resident staff for their joint work, and we dedicate this bit of real earth, its sprouting plants and its breeding animals, here and now to the study of the laws of the evolution of organic beings."

Davenport was followed on the program by Walter R.T. Jones, Governor of the Wawepex Society, who gave an entertaining Address of Presentation (see Prologue) in which he outlined the history of the Jones lands at Cold Spring Harbor on which the new Station would be erected. Youngest brother of Bio Lab founder John D. Jones, W.R.T. Jones reminisced that "in my early days the village, especially on the west side, showed its activity by noises from the continued hammering of iron, the resounding echo from the coopering shops, the clanging of boat-builders, and the buzzing of saws. When this business became no

(1) Plot plan of the campus of the Carnegie Institution Station for Experimental Evolution.

Published in the *Carnegie Institution of Washington Yearbook* for 1904, the plan shows the proposed site of the "Station building," later called the Main Building (now Carnegie Library) in relation to the existing director's "Residence" (now Davenport House), both directly west of "Cold Spring Creek" at the head of the harbor. To the north are the buildings of the Brooklyn Institute of Arts and Sciences, including the "Biological laboratory" (later called Jones Laboratory) and to its south the former whaling era warehouse (not identified on this map) later known as Wawepex Building.

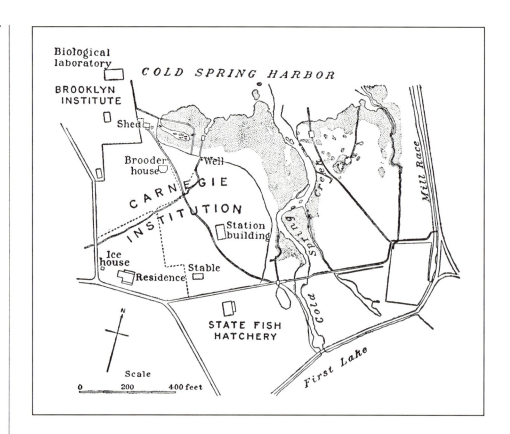

longer profitable, the place soon appeared like a deserted village—houses became vacant, buildings unused, and everywhere neglect and decay."

Carnegie Institution board chairman John S. Billings then gave the Response in which he expressed gratitude to the Wawepex Society for offering the use of their land (a total of ten acres) to establish the Station. Several buildings that stood on the Wawepex land were also put at the disposal of the scientists. These included the large house in which the Davenports had already taken up residence, a nearby stables and ice house, and a wharf and several harbor-side structures. Davenport would later describe these latter structures in great detail in his first Report to the Carnegie Institution on the new Station. *(1)*

> ...On the wharf there stands a shed, very useful for the temporary shelter of lumber, coal, etc., brought to the Station by boat. Just east is a large salt-water fish-pond, and beyond is a small boat and bath house, near which ways will lead to a larger boat-house for the protection of the Station launch during the winter. Near this boat-house and inside the main inclosure is a driven well 204 feet deep, flowing 9 gallons per minute. This will supply the residence, stable, and laboratory, by means of an electric pump with a capacity of 15 gallon per minute. It is proposed to supply the tanks in the cellar and first floor of the laboratory [not yet built] from a spring in the ravine... .

As the dedication ceremonies proceeded Franklin W. Hooper, representing the Brooklyn Institute of Arts and Sciences, delivered an official Welcome to the scientific newcomers to Cold Spring Harbor. Renowned Mendelian geneticist Hugo de Vries, the featured speaker of the afternoon, then took the platform. Professor de Vries, who had traveled to Cold Spring Harbor from the University of Amsterdam where he was director of the Botanic Garden, concluded his remarks as follows:

> Ladies and gentlemen, it is a high honor for me that this laboratory has been founded, and that the members of the board and the director have invited me to be its godfather. During a long series of years I have fostered my conception of sudden mutability and cultivated my primroses for myself and for myself only. Nobody knew about them.
>
> Some years ago I allowed myself to be induced to betray my secret and to deliver it to the scientific world. It has at once been taken up by your countrymen, and the foundation of this laboratory is the mightiest and the most dreadful competition that I could have. ...I have to submit to the prospect of being soon surpassed and largely excelled on the path which until now I considered as my own. I have to yield my much beloved child. But I do it gladly and without regret. It is the interest of the child itself which commands me. It will be better in your hands, Mr. and Mrs. Davenport, and in yours, lady and gentleman officers of the staff. Pray have good care of it and educate it assiduously, that it may become one of the most brilliant parts of your work, a glory to this laboratory and to the institution that founded it, a pride to our country, and a bliss for humanity.

The Carnegie staff will prove Mendel's laws through experiments in plant and animal breeding

The Carnegie Institution's Department of Experimental Evolution headquartered at the Station for Experimental Evolution in Cold Spring Harbor was to have four categories of members: Resident Scientists at the Station on Long Island; Honorary Associates, of which Professor de Vries was the first; Associates, nonresident scientists who received individual Carnegie research grants; and Correspondents, biologists anywhere in the world who wished to exchange comments and ideas on work in the field of experimental breeding. Even though at the time of the dedication the Station did not yet have its own facilities, Director Davenport had already appointed a full-time staff of eight, of which five were professional scientists. Buildings for conducting animal and plant breeding experiments soon began to go up. These structures eventually included numerous specialized animal and plant facilities and, most notably, two brick-trimmed, stuccoed laboratories—the Main Building and the Animal House. *(2)*

MAIN BUILDING
(Now Carnegie Library)
Kirby, Petit & Green
1905

The first research laboratory erected by the Carnegie Institution at Cold Spring Harbor was ready for occupancy early in 1905 and soon became known as the Main Building. This two-story rectangular structure built near the head of the harbor was a Mediterranean variant of the Second Renaissance Revival style (the first revival had come into vogue before the Civil War). Constructed of brick with frame partitions and floors and finished in stucco, it featured classical detailing around the doors and windows set off by brick trim. Its flattened hipped roof had flaring overhanging eaves supported by wooden brackets. A smaller monitor roof, also hipped, sat on top of the main roof. Between this higher roof and the main roof were short, horizontal clerestory windows to light the attic. *(3,4)*

In its general appearance the new laboratory for the Carnegie Institution could easily have been mistaken for one of the hundreds of libraries that the Carnegie Foundation erected for communities throughout the United States in the early years of the twentieth century. In fact, the Main Building is no longer used as a laboratory but, by happy coincidence, now houses the main library of Cold Spring Harbor Laboratory. Since the time of this transformation (1953) it has been known as Carnegie Library. *(5,6)*

The design of the building was commissioned from Kirby, Petit & Green, a New York City architectural firm proficient in all the popular styles of the day. Their American Bank Note Company Building, still standing at 90 Broad Street in Manhattan, was an impressive five-story limestone building in Classical Revival style featuring three-story-tall Corinthian columns. On Long Island the firm was responsible for the

(3) Main Building of the Carnegie Station (now Carnegie Library) as completed in 1905.

Published in *American Architect and Building News* (Vol. 88, No. 1541 [July 8, 1905]), this plate shows the new laboratory building in its pristine, symmetrical Second Renaissance Revival style, with brick-trimmed quoins on its corners and a classical arcade enframing the main entrance. Both the flaring main roof and the roof of the monitor section above it were sheathed in standing-seamed terneplate (steel coated with an alloy of lead and tin; no longer extant) giving them a Mediterranean appearance, as though they were covered in rows of clay pantiles.

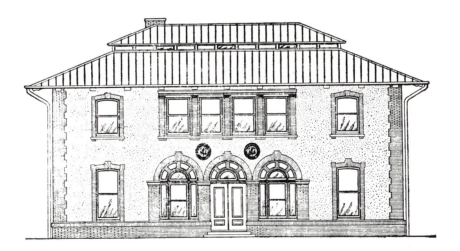

(4) Architectural rendering of the Main Building published in the *Carnegie Institution of Washington Yearbook* for 1904.

(5) Main Building today,
now Carnegie Library.

(6) Main Reading Room of Carnegie Library today.

neo-Tudor design of the Country Life Press Building in Garden City. Kirby, Petit & Green also designed consecutively two country residences for the founder of Country Life Press, Frank N. Doubleday, on the same site in Mill Neck, which is not far from Cold Spring Harbor. The first residence was designed in the picturesque shingled Colonial Revival style and the second in the more formal brick Georgian Revival style.

The Carnegie Institution's Italian Renaissance-inspired Main Building housed diverse but important functions at the Station. Two large Breeding Rooms occupied the south and east sides of the ground floor, and an Aquatic Animals room filled with aquaria was situated on the north side. An entrance Vestibule was located in the center of the front of the building. On either side

of the Vestibule were a Work Room and a Food Room, with a wide stair hall behind. *(7,8,9)*

Situated around the stair hall on the second floor was a series of rooms for Research, together with a Secretary's Room and a small Library. The entire south end was taken up by a Bird & Insect Room, which was lit from above by an extensive multipaned skylight (now gone).

On the unfenestrated west side of the basement floor were a Photographic Dark Room; a Dark Room for Cave Studies, to which a newly excavated cave was later connected; and a Low Temperature Room. On the east side, which had small windows facing in the direction of the harbor, were a Coal Room, a Boiler Work Room, and a room for Food Storage.

(7) Floor plans of the Main Building, 1904.

These plans published in the *Carnegie Institution of Washington Yearbook* for 1904 show uses that have now been superseded by the demands of state-of-the-art informational science. Serving today as the main library of Cold Spring Harbor Laboratory, the building houses photocopying and on-line computer services in the former cellar area (*top*), together with the main reference collection of books catalogued by the Library of Congress (the LC Collection) and quiet areas for study and writing. On the first floor (*middle*) the "Aquatic Animals" room has been converted to the Main Reading Room (see illustration 6) which contains current scientific journals and series (365 titles), and the former "Breeding Rooms" are now the Bound Journals Room. The library staff functions mainly out of the former "Work Room" and "Food Room." Much of the second floor (*bottom*), minus the "Sky-Light" that originally illuminated the "Bird & Insect Room," is now devoted to editorial offices for three journals published by Cold Spring Harbor Laboratory Press: *Genes & Development, Cancer Cells,* and *PCR Methods and Applications.* The clerestory-lit attic (plan not shown) under the monitor roof houses the Laboratory's Archives.

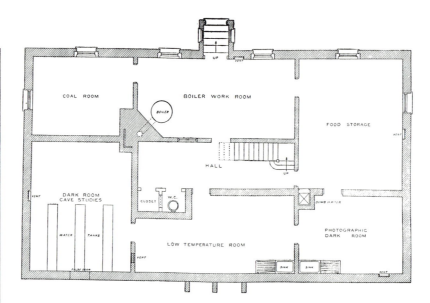

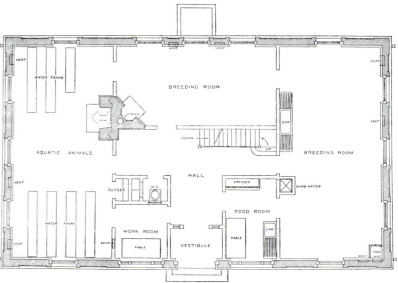

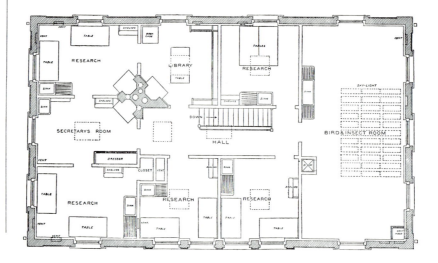

(8) Breeding Room of the Main Building in 1906.

According to the caption for this photograph as it appeared in the *Carnegie Institution of Washington Yearbook* for 1906 the room was "devoted to breeding canaries and other cage birds. There are about 175 birds in the room."

(9) Breeding Room in the 1920s.

The room is arranged for a reception in honor of the 25th anniversary of the founding of the Carnegie Station. At this time it was no longer used as a breeding room, the animals having been removed after the Animal House (now McClintock Laboratory) was completed in 1914 (but note the exhibit case on the left).

(10) George Shull's cornfield.

In this experimental garden to the north and east of the Main Building Shull raised inbred strains of maize (Indian corn). By strictly controlling their pollination—performing cross-pollinations by hand and not allowing the various strains to self-pollinate—in 1908 he demonstrated for the first time the phenomenon of "hybrid vigor" which increased seed yields by 20%.

George Shull, maize geneticist, demonstrates "hybrid vigor" in the cornfield behind the Main Building

Although the staff of the Carnegie Institution's Station for Experimental Evolution was small at first, important early work was performed both in and around the Main Building. Immediately after receiving his Ph.D. from the University of Chicago in 1906 George Shull (1874–1954) came to Cold Spring Harbor and began planting maize (Indian corn) in an experimental garden on the east side of the Main Building. By 1908 he was reporting in the *Carnegie Institution of Washington Yearbook* experiments that became world famous. Shull showed that when two different but carefully inbred strains of corn were crossed, the yield was 20% higher than if each strain were allowed to self-pollinate. This phenomenon of "hybrid vigor" that he demonstrated experimentally was later employed in commercial seed production to create high-yielding strains that today make corn the most important agricultural crop in the United States. (10)

Brooder houses and other specialized animal buildings are erected

While George Shull worked with corn, Station director Charles Davenport performed breeding experiments with a variety of domestic animals, including cats, chickens, dogs, goats, and sheep, in the hope of demonstrating Mendelian patterns of inheritance

(11) "View across Cold Spring valley looking south-eastward, showing part of the grounds of the Station for Experimental Evolution."

This caption for the photograph as it appeared in the *Carnegie Institution of Washington Yearbook* for 1906 continues as follows:

"Main building at the extreme right, potting house and propagation house in front, and vivarium, under construction, in front of and to left of latter. To the left (north) of the main building is seen part of the east experimental garden. Near the extreme left is the brooder house, from which radiate eight poultry runs, seen in the middle foreground."

among his experimental animals. Specialized buildings such as aviaries, brooder houses, chicken coops, poultry runs, and a sheep shed were erected as needed. Apparently some of the animals were housed in the Main Building itself, as the original names for the rooms suggest, but this was not to be a long-term solution. *(11, 12)*

The Report of the Department of Experimental Evolution for the year 1911 revealed that "it was finally decided to move all breeding animals to a new building, relieving the Main Building of the dirt that is inseparable from their culture and allowing expansion in it of the space available for records and their study." The Blackford Hall architect Robert W. Gardner (see Chapter One) was commissioned to design a new building "that could be used for chemical studies on mammals and for operating upon them."

(12) "View of part of sheep and goat pasture and poultry runs. Brooder house to the left."

Published in the *Carnegie Institution of Washington Yearbook* for 1906, this view also shows the Station's storage shed and wharf in the background.

Only a small portion of the proposed building was completed in 1911 to Gardner's specifications. Whether this was due to a lack of funds, as stated in the Report, or to dissatisfaction with Gardner's design is not entirely clear. Presumably the design called for reinforced concrete, and the structure may have been too utilitarian in appearance for Director Davenport's liking, even for an animal facility. A year later there were still reports of overcrowding in the Main Building but no indication of progress on the building designed by Gardner. In the meantime a researcher on pigeons had been hired, and his birds also had to be housed, or as Davenport tersely phrased it, "new needs arise with the development of our science which require additional equipment." Gardner was let go, and the New York City architectural firm of Peabody, Wilson & Brown was entrusted with completion of the building. *(13,14)*

(13) Plans of the Animal House.

As published in the *Carnegie Institution of Washington Yearbook* for 1913 (the year before the building was completed) these plans indicate as "Present" (lower left-hand corner of the building) the section previously completed in 1911 to the design of Robert W. Gardner, the architect of Blackford Hall. He was succeeded by the architectural firm of Peabody, Wilson & Brown who produced these plans for a considerably larger building.

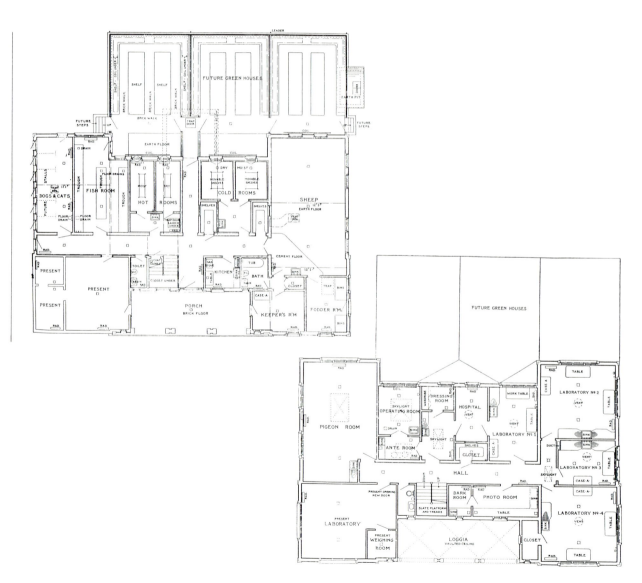

EXPERIMENTAL RESEARCH BUILDING, CARNEGIE INSTITUTE, COLD SPRING HARBOR, L. I.

Peabody, Wilson & Brown, Architects.

(14) Main elevation and plan of the Animal House (now McClintock Laboratory).

This plate reproduced from *Architecture* (Vol. 29–30 [1914], p. 221) shows off the classical lines of this "Experimental Research Building" to great advantage, particularly its arcaded entrance, brick-trimmed quoins, and parapetted roof line. As the room designations suggest (e.g., "Rabbits," "Fish," "Sheep,"), the animal breeding experiments at Cold Spring Harbor were moved from the Main Building to this new facility upon its completion in 1914.

(15) Animal House today, now called McClintock Laboratory.

The building was extensively renovated in 1971 and then renamed in honor of Carnegie researcher and Nobel laureate Barbara McClintock.

ANIMAL HOUSE
(Now McClintock Laboratory)
Peabody, Wilson & Brown
1914

The new Carnegie facility (finally completed in 1914 and soon known as the Animal House) was a two-story stuccoed building with hollow tile walls and concrete floors. Like the Main Building, its classical features were detailed in brick, and it could be called Second Renaissance Revival in style, but with a difference. It appears to have been patterned after a famous real-life building in Italy, the historic Stazione Zoologica in Naples, and the suggestion of this model most likely came from Director Davenport.

Built in 1872 on a site overlooking the Bay of Naples, the Stazione Zoologica was the first permanent seaside biological laboratory in modern times. In 1902 Davenport made a tour of European biological laboratories on a fact-finding mission aimed at bolstering his chances of winning the support of the Carnegie Institution for a comparable station at Cold Spring Harbor. Undoubtedly he would have visited the Naples station and could easily have furnished the architects with a drawing or photograph of the building. Davenport may also have been influenced in his (putative) choice of a model for the new laboratory by the fact that his mentor at the University of Chicago, Charles Otis Whitman, did his early research at the Naples station.

Whether or not Davenport chose the architectural style to be used, the Animal House at Cold Spring Harbor and the Stazione Zoologica at Naples in fact shared several features in common. Besides having hipped roofs, similar oblong shapes, and almost identical fenestration patterns, the feature that most made them look alike was the way the center part of the main facade was recessed behind a tall arcaded portico. *(15,16,17)*

The ground floor of the Animal House was devoted to rooms for the housing and care of various animals, including a room for Dogs & Cats, a Fish Room, and an earth-floored room for Sheep, plus a Keeper's Room, Fodder Room, Hot Rooms, Cold Rooms, and Kitchen. A spacious Pigeon Room with a large skylight (now gone) was on the floor above, along with three large Laboratory rooms, a Dark Room, a Photo Room, and two smaller Laboratory rooms, one of these ensuite with a Hospital, Dressing Room, and Operating Room. The plans of the Animal House (see illustration *13*) published in the *Carnegie Institution of Washington Yearbook* for 1913 show Future Green Houses abutting the rear facade of the structure, but these were not erected in that location.

When Davenport entrusted Peabody, Wilson & Brown with the design of the new facility the firm was busy with many commissions on the north shore of Long Island. The majority of these buildings were executed in the Colonial Revival or Georgian Revival style, usually with brick facades. Surviving local examples include the former Huntington Town Hall, 1912

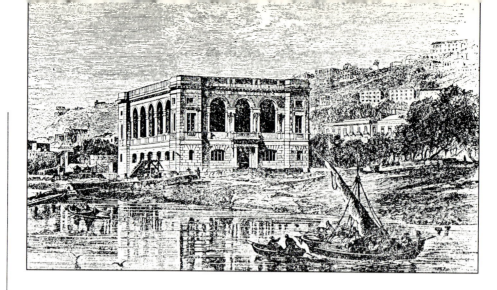

(16) Naples Stazione Zoologica (Zoological Station) which probably served as the model for the Animal House.

Built overlooking the Bay of Naples in 1872, this biological sciences training institute was much visited by Americans as well as by Europeans. Undoubtedly it was on Charles Davenport's itinerary when he visited European facilities in 1902. He could easily have supplied a likeness of this building to the architects Peabody, Wilson & Brown whom he commissioned with the final design of the Animal House for the Carnegie Station.

(now Sammis Realty), the former Cold Spring Harbor Library, 1913 (now the Gallery of the Society for the Preservation of Long Island Antiquities), and the Union Free School in Cold Spring Harbor, 1925 (now the DNA Learning Center of Cold Spring Harbor Laboratory). In addition to numerous large country estates, other Long Island commissions included the Children's Library and the Robert Mason Memorial Children's Hospital, both in Westbury.

(17) Animal House in the 1920s.

Free-standing aviaries are visible in the background on the far left.

(18) Power House in the 1960s.

Built in 1913, the Power House (*center*) was erected to simplify heating all of the Carnegie Institution buildings at Cold Spring Harbor, including greenhouses and aviaries as well as the two laboratory structures: the Main Building (its roof visible in the background on the left) and the Animal House (then still under construction). The Power House had three rooms: a Coal Bunker, a Boiler Room, and a Machine Room. The building directly to the south (*left*) is the Carpentry Shed. A portion of the Bio Lab's Jones Laboratory is visible in the foreground on the right.

POWER HOUSE
Peabody, Wilson & Brown
1913

As the Animal House neared completion a Power House was quickly constructed directly to the north to facilitate heating all of the Carnegie structures (including aviaries and greenhouses built to the west and north of the new Animal House). Also designed by Peabody, Wilson & Brown, the story-and-a-half Power House completed in 1913 was built of hollow tile and concrete and finished in stucco. Originally it contained a Boiler Room, Coal Bunker, Machine Room for pumps and electric generators, and a Forge Shop. Today the ground floor is used by Cold Spring Harbor Laboratory's Buildings and Grounds Department for equipment storage, and the attic story, lit by new skylight windows, houses the Buildings and Grounds administrative and design offices and the Health and Environmental Safety Office. *(18, 19)*

(19) Power House today.

The Power House still contains an emergency generator and houses much of the equipment required by Cold Spring Harbor Laboratory's Buildings and Grounds Department, whose administrative offices are now located on the second floor, together with the Laboratory's Health and Environmental Safety Office.

The Davenports look into the inheritance of eye, hair, and skin color

In addition to writing theoretical papers on the methods and results of pedigree breeding of plants and animals (published in the journals of the American Philosophical Society and the Society for Experimental Biology and Medicine) and pursuing experimental work on canaries, poultry, and sheep (published by the American Breeders Society), in 1907 Davenport began studies on the inheritance of human physical characteristics, an endeavor in which he was ably assisted by his wife. Together the Davenports pursued research into the genetics of eye, hair, and skin color and coauthored a number of scientific papers between 1907 and 1910 (published in *Science* and *The American Naturalist*) on such subjects as "Heredity of Eye-color in Man," "Heredity of Hair-form in Man," "Heredity of Hair Color in Man," and "Heredity of Skin Pigmentation in Man."

Charles Davenport reports on eugenics and inaugurates a summer course on the subject

Davenport discussed human heredity in a social rather than a strictly experimental sense for the first time in two short papers that appeared in 1909: "Family Records" (privately printed) and "Influence of Heredity on Human Society" (published in the *Annals of the American Academy of Political and Social Science*). The following year his small book entitled *Eugenics: The Science of Human Improvement by Better Breeding* (Henry Holt, 1910) was published, and it may have been used as a text for the new Eugenics course first offered at Cold

Spring Harbor in the summer of that same year. The word "eugenics" (from a Greek root meaning "good in birth") was coined by a cousin of Charles Darwin, the English anthropologist Francis Galton, and is defined as "a science that deals with the improvement (as by control of human mating) of hereditary qualities of a race or breed." The Eugenics course at Cold Spring Harbor, which was taught by Davenport with the assistance of former Bio Lab pupil and agricultural breeder Harry H. Laughlin, trained workers in the techniques of recording and compiling family pedigrees in an effort to understand patterns of human inheritance.

Davenport founds a Eugenics Record Office
(ERO) near the Bio Lab with the help of
Mrs. E.H. Harriman

Over the years the Eugenics course enrolled far greater numbers of students than any of the basic science courses at the Bio Lab. This was possible only because Davenport had early on managed to obtain the use of a nearby residence for housing the students, administering the pedigree compilations, and storing the records that the eugenics field-workers trained at Cold Spring Harbor would amass in the future. Firmly committed to establishing an important eugenics research center in conjunction with the ongoing work of the Bio Lab and the Carnegie Station, Davenport had located a potential "angel" in the person of Mary Williamson Harriman, who had been recently widowed by railway magnate Edward Henry Harriman. The knowledge that her daughter Mary, with a B.A. from Barnard College, had been enrolled in the 1906 Embryology course bolstered Davenport's confidence as he pursued Mrs. Harriman, and his efforts met with immediate and great success. To provide facilities for the proposed eugenics program Mrs. Harriman purchased a mid-Victorian residence situated nearby on 75 acres to the east of Stewart Lane and north of New York State Route 25A in Laurel Hollow (just to the west of the property under the jurisdiction of the Carnegie Institution). Here, under the auspices of the American Breeders' Association, a Eugenics Record Office (ERO) was established in the fall of 1910.

The Stewart House becomes the first headquarters
of the ERO (today it is once again a private residence)

The house on the newly acquired ERO property had been built shortly after the 1839 marriage of Manhattanite Charles P. Stewart to Helen Jones, youngest daughter of Cold Spring Whaling Company founder John H. Jones. After it became the ERO head-

(20) First home of Eugenics Record Office (formerly the Charles Stewart House and today once again a private residence).

Built in the mid-nineteenth century for gentleman farmer Charles Piers Stewart, who had married into the Cold Spring Harbor Jones family, this Italianate-style home, together with 75 acres, was purchased in 1910 to house the activities of the newly established Eugenics Record Office (ERO).

quarters the large clapboarded Stewart residence gained a masonry wing at its north end to provide fireproof storage for the eugenics records. The students who came in the summer to prepare for eugenics fieldwork literally pitched their tents on the grounds of the residence (tents later became popular at the Bio Lab campus as well). Their training involved many field trips to examine records at hospitals and prisons as well as to conduct door-to-door surveys. *(20,21)*

(21) Tents on the grounds of the Eugenics Record Office.

The students in the Eugenics course were housed at first in wood-framed canvas tents pitched in the sheep meadow of the Stewart farm.

Mrs. Harriman makes a gift of a new ERO office
(a private residence today)

It was not long before a larger and entirely fireproof building was needed to house the records being diligently compiled by the eugenics workers. Mrs. Harriman again rose to the occasion by donating $15,000 to erect and furnish a purpose-built two-story brick and stucco structure. Designed in Second Renaissance Revival style by Peabody, Wilson & Brown, the architects of the Animal House, and completed in the same year, 1914, the new Eugenics Record Office building contained offices for the ERO staff (Superintendent Laughlin, an archivist, an editorial secretary, and four clerks) as well as plenty of space for the storage and study of the eugenics records. After the new structure was completed the Stewart House became a residence hall for ERO staff and students. *(22,23)*

With the long-term future of the Eugenics Record Office in mind, Mrs. Harriman deeded the entire ERO property to the Carnegie Institution in 1914. Her subsequent gift of $300,000 enabled the Carnegie Institution to endow a Department of Genetics at Cold

EUGENICS RECORD OFFICE, CARNEGIE INSTITUTE, COLD SPRING HARBOR, L. I.   Peabody, Wilson & Brown, Architects.

(22) New Eugenics Record Office building erected in 1914 (now privately owned).

As shown in this plate from *Architecture* (Vol. 29–30 [1914], p. 222) the new building for the Eugenics Record Office (designed by the architects of the Animal House, Peabody, Wilson & Brown) was a brick-trimmed stuccoed structure in Second Renaissance Revival style. It was built on the east side of Stewart Lane immediately to the south of the Charles Stewart House, the ERO's first headquarters. The ERO was closed in 1940, and its buildings and lands were subsequently sold by the Carnegie Institution.

(23) Eugenics Record Office interior with files.

In this 1920s view of an airy and bright workroom located on the second floor of the new ERO building, eugenics workers are studying records such as family pedigrees that were obtained on visits to institutions and by door-to-door canvassing. The volumes on the shelves included many genealogical works as well as journals on the subjects of breeding, heredity, and eugenics—"a science that deals with the improvement...of hereditary qualities of a race or breed."

Spring Harbor in 1921. This was an amalgamation of the eugenics studies begun at the ERO in 1910 with the program of experimental breeding of animals and plants inaugurated on the grounds of the Station for Experimental Evolution in 1904.

Davenport leads the American eugenics movement

In the second year that the Eugenics course was taught Davenport published a textbook entitled *Heredity in Relation to Eugenics* (1911). This was his most influential publication in the field of eugenics and established him as the major figure in the American eugenics movement. However, as noted in a biographical memoir written in 1944 by his long-time associate E. Carleton MacDowell (published by the Carnegie Institution under the title "Charles Benedict Davenport, 1866–1944: A Study of Conflicting Influences"), although Davenport's famous book bore witness to a wide range of putative human hereditary traits, "its continued usefulness was reduced by its hasty preparation and the lack of critical judgment in lumping together...cases with ample and insignificant evidence. The topics of Davenport's special studies covered a wide range...[including] stature, body build and longevity, goiter, pellagra, epilepsy, temperament, mongoloid dwarfs, twinning, sex-linkage, and race crossing." He even identified a trait which he called "thallassophilia" (love of the sea) that supposedly ran in families that produced high numbers of naval officers.

In contrast to most geneticists of his time Davenport seemed to believe that everything that "ran in families" must be hereditary. As noted by Stony Brook geneticist and science historian Bentley Glass (in a work in progress), "every geneticist, even of that time, knew of the profound effects of the environment on agricultural plants and domestic animals." Glass also points out that what Davenport failed to consider was "how to calculate the influence upon the development of human characteristics of all sorts of possible environmental influences, such as health, nutrition, schooling, poverty, indoctrination, and many other factors." Commenting on the memoir by MacDowell, Glass concludes that it contains the "objective criticism of a younger colleague who admired the older man greatly, but was not blind to his failure to do good scientific work." The real tragedy of the situation, as MacDowell himself wrote, was that Davenport's "scientific background and associations gave the prestige of an authority in the eyes of those inclined to accept his position on the social and political aspects of eugenics. But the opposition of many [fellow geneticists], instead of quickening the research for more accurate and convincing evidence, called forth a defensive attitude which led to exaggerated emphasis and dulled objective thinking."

Davenport's worst error of judgment was his failure to disassociate himself from the overt racism of Henry Laughlin, the man whom he had appointed to direct the Eugenics Record Office at Cold Spring Harbor in 1910. Laughlin later made himself out to be an authority on the "biological" side of the immigration issue, and his testimony on Capitol Hill helped to pass not only federal legislation (the 1924 Johnson Act) barring southern and eastern Europeans from immigration to the United States but also many state laws requiring forced sterilization. It was not until 1939 that the Carnegie Institution finally persuaded Laughlin to retire from his ERO post, and the following year the Eugenics Record Office closed its doors, the Carnegie support having been completely withdrawn.

"It is the business of the Department of Experimental Evolution to study the behavior of this germ-plasm...."

As the Carnegie Station director Davenport early on showed himself to be the intellectual heir of the original eugenicist Francis Galton, sharing his belief that questions of inheritance could be solved by the methods of biometry—"the statistical analysis of biological observations and phenomena." In particular Davenport believed that by applying statistical methods to the study of evolution he and his colleagues at the Station in Cold Spring Harbor could obtain definitive proof for Darwin's theory of natural selection, which at the turn of the century was still not accepted by many members of the scientific community. Such a biometrical approach had already guided the brilliant corn breeding experiments of George Shull, one of Davenport's first appointments to the Station for Experimental Evolution. By way of preface to his Report to the Carnegie Institution in 1908 (the same year that Shull published his discovery of hybrid vigor in corn), Davenport noted that "certain portions of that unending stream of reproductive matter which has come down to us from the time when life began on earth and by changes in which all evolution has taken place are now under our careful observation and to a large extent under our control. It is the business of the Department of Experimental Evolution to study the behavior of this germ-plasm and to note its reaction to the conditions we impose." The scientists Davenport enlisted to undertake this task did so with great enthusiasm, but their prospects for experimental success would be hampered in some cases by the limited extent to which Mendelism could explain patterns of inheritance.

In their ongoing work to try to effect genetic changes by experimental means they employed a wide variety of techniques on an eclectic assortment of organisms. In the basement of the Main Building, for example, starting in 1908 Charles Banta conducted cave experiments aimed at inducing color mutations in tadpoles of the tiger salamander. E. Carleton MacDowell, fresh from obtaining his Ph.D. at Harvard in 1914, was given the task of seeing whether alcohol could cause gene mutation; if so, the tragic lives of families with long histories of alcohol abuse might be explained as being the result of a degeneration of their germ plasm. After several years of negative results MacDowell abandoned this project. In time the research at the Station for Experimental Evolution became focused on four animal and plant subjects—the mouse, the jimson weed, the fruitfly, and the pigeon—whose names were to become intertwined for scientific posterity with those of their Cold Spring Harbor investigators—MacDowell and Little, Blakeslee, Bridges and Demerec, and Riddle.

C.C. Little and Carleton MacDowell discover genetic components of cancer in mice

Clarence Cook Little (known as C.C. Little; 1888–1971) was in residence at Cold Spring Harbor for only four years, but during his second year (1920) he published a paper on "The Heredity and Susceptibility to a Transplantable Sarcoma of the Japanese Waltzing Mouse" in which he reported results demonstrating that resistance to cancer can differ between different strains of mice. Little served briefly as assistant director at Cold Spring Harbor (from 1921 to 1922) before moving on to the presidency of the University of Maine (1922–1925) and thereafter to that of the University of Michigan (1925–1929). In 1929 he moved permanently to Bar Harbor, Maine, where at the age of 42 he founded the Jackson Laboratory, now the largest supplier of purebred mice in the United States. Other mouse-oriented research at Cold Spring Harbor was performed by Carleton MacDowell. After his unsuccessful attempt to demonstrate a mutagenic (gene-altering) effect of alcohol on mice, MacDowell happened upon a mouse strain that always develops leukemia. He later developed strains of mice with varying degrees of resistance to leukemia. These strains, like Little's, resulted in the establishment of many of the mouse stocks upon which today's cancer research depends.

Albert Blakeslee demonstrates chemical mutagenesis
in the jimson weed

The first successful demonstration of chemical mutagenesis at Cold Spring Harbor occurred using a plant rather than an animal species. Albert F. Blakeslee (1874–1954), who joined the staff of the Carnegie Station in 1915, showed in the course of his work with the jimson weed *Datura stramonium* that the alkaloid drug colchicine can cause chromosome duplications. Blakeslee had visited Cold Spring Harbor many years before, having served for two summers (1901 and 1902) as an assistant in the Botany course taught at the Bio Lab, and just before returning to Long Island to take up the Carnegie position he taught one of the first modern genetics courses in a summer session at Storrs Agricultural College in Connecticut. At Cold Spring Harbor Blakeslee raised more than 70,000 *Datura* plants each summer between 1915 and 1941 in an experimental garden at the water's edge, just east of the Animal House. Besides studying inheritance in the jimson weed, he also explored the genetics of taste and smell and proposed as an example of a simple recessive characteristic in humans the inability to taste PTC (phenylthiocarbamide), which is extremely bitter to those who can taste it. *(24,25)*

(24) Albert Blakeslee's *Datura* seedlings.

This view of one of the Carnegie greenhouses shows seed pans and seedlings of *Datura stramonium,* or jimson weed, the plant in which Carnegie researcher Albert Blakeslee induced chemical mutagenesis using the alkaloid colchicine, which caused chromosome duplications in his plants.

(25) Blakeslee's field of *Datura* plants.

For over twenty-five years, starting in 1915, Blakeslee planted upwards of 70,000 *Datura* seedlings each summer in his experimental garden to the east of the Animal House. At the near left is the Carpentry Shed and at the far left is the former Bungtown Barrel Factory. The Sand Spit stretches across the photograph in the distance.

Blakeslee first became active in the administration of the Carnegie Department of Genetics in 1923 when he was made assistant director. He later became director upon the retirement of Charles Davenport in 1934. By this time Davenport was backing away from eugenics matters and he increasingly devoted his retirement years to the affairs of the fledgling Whaling Museum in Cold Spring Harbor, which he helped found in 1936. Davenport's devotion to the Whaling Museum would tragically lead to his death in 1944. After laboring day and night in his backyard in the winter's cold boiling the remains of the head of a beached whale to retrieve its bones for a new exhibit at the Museum (still on view today) he contracted pneumonia and died shortly thereafter at age 78.

Milislav Demerec and Calvin Bridges map the fruit fly chromosomes

Blakeslee's first appointment in his capacity as assistant director in 1923 had been Cornell-trained cytogeneticist Milislav Demerec (1895–1966). Although trained as a maize geneticist, Demerec soon turned his efforts at Cold Spring Harbor to the fruit fly *Drosophila*. The chromosomes of the fruit fly were first studied here by Charles William Metz, a

Carnegie Institution investigator from 1914 to 1930, and later by Calvin Bridges, who first visited Cold Spring Harbor as a Bio Lab research fellow in the summer of 1919. In the late 1930s Demerec and Bridges, who was then at the California Institute of Technology, collaborated during the summers on drawing large-scale maps of the four *Drosophila* chromosomes, which was considered quite a feat in those days. By 1940 the first edition of Demerec's *Drosophila Guide* appeared. This classic in its field was destined to go through eight editions, the last appearing in 1969. *(26)*

Oscar Riddle, the pigeon man, isolates the lactation-inducing hormone

Oscar Riddle (1877–1968) arrived at Cold Spring Harbor before Blakeslee, Demerec, and Little and, except for Demerec, remained for the longest time (1914–1945). Davenport had brought Riddle to the Carnegie Station to carry forward the life's work of University of Chicago professor Charles Otis Whitman (1844–1910), who had been mentor to both men. Whitman had achieved great renown for his studies on the anatomy and be-

(26) A *Drosophila* laboratory. This view of a room on the second floor of the Main Building (now Carnegie Library) shows items needed in studying *Drosophila*. The tiny fruit flies were fed a diet of mashed bananas and were bred in the stoppered milk bottles seen on the shelves in the foreground. On the work tables are vials of chloroform, lamps, and microscopes used for anesthetizing the flies and examining them for mutations.

(27) Oscar Riddle's aviaries behind the Animal House.

This 1920s photograph shows the large amount of space devoted to the care of the Carnegie Institution's breeding flock of pigeons.

(28) Riddle's metabolism laboratory in the Animal House.

Riddle was the world's expert on the morphology and physiology of pigeons. His proficiency in biochemistry far surpassed that of all of the other workers at Cold Spring Harbor who collaborated with him to their great advantage in isolating and characterizing "internal secretions" (hormones) in the 1930s.

havior of pigeons, a large group of which he bred, cared for, and studied on the grounds of, and even inside, his own home. Whitman's other claim to fame was as founder and first director of the Marine Biological Laboratory at Woods Hole, Massachusetts, the seaside zoological station established in 1888 for much the same purposes as the later Biological Laboratory at Cold Spring Harbor.

When Whitman died in 1910 he bequeathed his entire breeding stock of pigeons to Oscar Riddle, and these birds then became Riddle's lifetime work. This was the flock for which the Pigeon Room on the second floor of the Animal House was designed. When Whitman's breeding colony inevitably outgrew these quarters, a sprawling complex of pigeon houses took shape on the hill directly behind the Animal House. Riddle's first responsibility upon arrival in Cold Spring Harbor, however, was the task of editing Whitman's voluminous lifelong research notes, which covered every known scientific discovery regarding "natural and hybrid pigeons and doves." These edited notes eventually appeared in three volumes in 1919 under the Carnegie Institution of Washington imprint. *(27)*

Riddle's interest in the "internal secretions" (later called hormones) of pigeons led him to a collaboration with rabbit researchers working along similar lines. Partly as a result of his familiarity with biochemical techniques acquired over nearly two decades of pigeon work, Riddle was the member of the Cold Spring Harbor team who performed the final steps in the 1932 isolation of the hormone prolactin from the rabbit anterior pituitary gland, and it was he who demonstrated that the purified extract was indeed the protein substance that produces lactation in warm-blooded animals, including the secretion of "milk" in the crops of pigeons. Riddle's hormone work in the 1930s, so different from the earlier work of his colleagues in the Carnegie Department of Genetics, was closer in spirit to the research then being pursued on a new year-round basis at the neighboring Biological Laboratory. In fact, his collaborators in this work were recent appointees to the Bio Lab. *(28)*

CHAPTER THREE

LIBA, a group of neighbors, takes charge of the Bio Lab

1920~1940

Within ten years of its founding the Carnegie Station for Experimental Evolution had expanded both physically and scientifically to the point where it greatly overshadowed the neighboring Biological Laboratory, still in essence a summer training school for biology teachers. Far from expanding, the program at the Bio Lab had remained relatively static since the arrival of the Station in 1904. Interest in the elementary level courses offered at the Bio Lab had fallen off, and with the exception of the notoriously popular Eugenics course few new courses were added after the turn of the century. With the outbreak of hostilities in Europe in 1914 the number of enrollments at the Bio Lab dropped precipitously, and in the years following World War I its always tenuous relationship with the Brooklyn Institute was once again in question. Its ties to the Brooklyn Institute were finally severed in 1924 when the ever-supportive local community assumed control of the Bio Lab under the aegis of the Long Island Biological Association (LIBA). Over the next ten years the long dreamed of goal to establish a year-round research program became a reality as new laboratory and housing facilities were constructed or purchased. Research flourished amidst these new surroundings, and by the time of its 50th anniversary in 1940 the Bio Lab was internationally recognized for both its research and its annual Symposium begun in 1933 by Reginald Harris.

Explaining Life in Terms of Molecules

Long into the first part of the twentieth century biologists belonged to two camps: the mechanists (reductionists), who argued that all aspects of life eventually could be explained in terms of the laws of physics and chemistry, and the vitalists, who believed that some still undiscovered special holistic principle underlies the living state. Prominent among the early mechanists was European-born Jacques Loeb, who spent much of his American career at the Rockefeller Institute and who used his summer General Physiology course at Woods Hole to combat mystical interpretations of living phenomena. In his 1906 book *The Dynamics of Living Matter* Loeb emphasized the importance of the recent demonstration by the German chemist Edward Buchner that the fermentative breakdown of sugar can occur outside of cells in yeast cell extracts. Perceptively Loeb wrote: "The history of this problem is instructive as it warns us against considering problems as beyond our reach because they have not yet found a solution." Starting from Buchner's seminal discovery, biologists trained in chemistry (who gradually began to call themselves "biochemists") showed over the next three decades that many other intracellular chemical reactions could occur outside cells, the key component being special catalytic proteins that came to be called enzymes.

Initially the cellular chemical reactions best understood in terms of physics and chemistry involved small molecules such as the sugars, fats (lipids), and amino acids. The science of how food molecules are broken down into intermediates, which in turn make new cellular building blocks, became known as "intermediary metabolism."

Many of the best scientific minds of the twentieth century were attracted to this research. Among them were the Berlin-based Otto Warburg and Carl and Gerty Cori, who came from Europe to eventually settle in St. Louis. Key to their success was the isolation of the enzymes that speed up (catalyze) their respective chemical reactions, and thus the designations biochemist and enzymologist became virtually interchangeable. Much less understood were the enzymes themselves. Invariably they were found to be proteins built up from many hundreds of amino acids. They were so large that they were known as macromolecules. How enzymes actually catalyze chemical reactions was not then a solvable problem. Working out the exact structure of even the smallest protein was beyond the abilities of any chemist of the time.

In addition to the chemists who worked on biological molecules, many scientists with physics backgrounds moved into biology. Often called "biophysicists," many concerned themselves with the electrical aspects of cells. They focused in particular on the outer membranes that surround all cells and control the selective movement of ions such as Na^+ and K^+ into and out of cells. By 1925 the Danish physicist Hugo Fricke, then working in Cleveland, had already deduced from electrical measurements the very thin 50-Å thickness of red blood cell membranes, and by the end of the 1930s the basic cell membrane structure was generally believed to be a protein-containing lipid bilayer.

Membrane structure was at the heart of understanding how signals are conducted along nerve cells. The inner surfaces of cell membranes are more negatively charged than the outer surfaces, and underlying these charge differences are differences in the concentrations of Na^+ and K^+ ions on the two sides of the cell membrane. Vital to the eventual understanding of how nerves work was the discovery in 1936 of the very large nerve fibers of the giant squid. Key biophysicists quickly moved to work with these fibers, and Kenneth Cole and Howard Curtis, then working at the College of Physicians & Surgeons of Columbia University, observed increases in ion movements during the propagation of signals along their experimental nerve fibers.

Other physicists came into biology to see whether X-ray crystallographic techniques could be used to work out the exact structures of proteins. To use this approach crystals must exist, and great excitement was generated when Irish-born J. Desmond Bernal and his student Dorothy Crawford (Hodgkin) working at the Cavendish Physics Laboratory of Cambridge University obtained in 1934 an X-ray diffraction photograph of the enzyme chymotrypsin. To go from X-ray photos to the three-dimensional structure of a protein, however, was then a more than herculean experimental objective. Not until 1959, in fact, did Max Perutz and John Kendrew at Cambridge achieve this goal while working on the oxygen-carrying proteins hemoglobin and myoglobin.

The Long Island Biological Association comes to the rescue

Only three years after the Bio Lab had fought for and won its rightful status as a separate department of the Brooklyn Institute of Arts and Sciences incredibly in 1920 the Brooklyn Institute once again entertained the notion of turning the Bio Lab over to another institution, this time to the University of Pittsburgh. To complicate matters further, many of the Bio Lab's friends mistakenly believed that the Bio Lab could somehow rely on the Carnegie Institution for support. This was emphatically denied by Director Davenport: "...it [the Bio Lab] actually receives and can hope to receive nothing from it [the Carnegie Institution]." When in 1921 Davenport announced his intention to step down from the directorship of the Bio Lab "at once," citing as the reason the now full-time demands of his position as director of the Carnegie Station, the future of the Bio Lab was far from secure.

Once again the Bio Lab's loyal supporters, especially the members of its Board of Managers, moved into action. A "Long Island Committee" was formed by one friend inviting another friend to participate in a drive not only to raise money for the Bio Lab, but also to help raise its profile in the community. Cast into the role of leading this drive was prominent local citizen Colonel Timothy S. Williams, president of the Brooklyn Rapid Transit System Companies. Upon his retirement from the Transit System in 1923 Williams focused his energies on heading up this now large and influential group of friends of the Bio Lab. That same year the Brooklyn Institute relinquished all future responsibility for the affairs of the Bio Lab to this group, which formally incorporated the following year as the Long Island Biological Association.

LIBA appoints Reginald G. Harris the new director

Even before LIBA's formal incorporation in 1924 a strong and relatively young Board of Directors was activated. The new LIBA board that Colonel Williams presided over included investment banker Marshall Field (only 31 years old) as vice president, and publisher, editor, and businessman Arthur Hines Page (just 40) as the first treasurer. Charles Davenport sat on the LIBA board both as secretary and as the representative of the Carnegie Institution of Washington. The Wawepex Society was represented by John H. Jones Stewart, grandson of John H. Jones and nephew of the Bio Lab's founder John D. Jones. Other early board members included Walter B. Jennings, William K. Vanderbilt, Jr., Mortimer L. Schiff, Henry W. de Forest, Childs Frick, and Russell Leffingwell. Most of these LIBA directors had homes in the Three Harbors area, which encompassed not only Cold Spring Harbor, but also

(1) Montage of North Shore mansions belonging to members of the Long Island Biological Association.

William J. Matheson, Fort Hill, Lloyd Harbor

Childs Frick, Clayton, Roslyn

Walter B. Jennings, Burrwood, Cold Spring Harbor

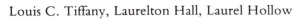

Louis C. Tiffany, Laurelton Hall, Laurel Hollow

William R. Coe, Planting Fields, Oyster Bay

Mortimer L. Schiff,
Northwood, Oyster Bay

William K. Vanderbilt, Jr.,
Eagle's Nest, Centerport

areas overlooking Lloyd Harbor and Huntington Harbor, both to the east. There was also strong representation from Oyster Bay and other harbor towns to the west. In addition to these neighbors, the LIBA board included distinguished men of science representing institutions such as Brown University, Columbia University, Cornell Medical College, Johns Hopkins University, Smith College, and Yale University. *(1)*

Davenport's resignation having been accepted, the LIBA board appointed his son-in-law Reginald Gordon Harris (1898–1936), a newly minted Ph.D., as the new director. Harris had married Jane Joralemon Davenport in 1922, the year before he received his advanced degree. He had earned both his B.A. and Ph.D. from Brown University, with advanced studies in Paris and anthropological and entomological field work in South America. Like many other biologists trained at Brown University, he had spent time at Cold Spring Harbor in the summer as a student at the Bio Lab and also as an assistant in the Carnegie Institution's Department of Genetics.

Colonel Timothy Williams raises the money to buy Jones land for new laboratories

When Reginald Harris became director of the Bio Lab it had few buildings of its own and barely three acres of land. The Carnegie Institution continually put up new structures on its more spacious campus next door to meet the needs of its growing genetics program at Cold Spring Harbor, but the last building erected at the Bio Lab had been Blackford Hall back in 1907. Although the Bio Lab enjoyed the use of a number of old residences and warehouses along Bungtown Road, the only building specifically designed for science was 1893 Jones Laboratory.

During the twenty-five years that Charles Davenport served as director of the Bio Lab he had many times in his writings expressed his belief that to survive the Bio Lab must expand. As the new director, Harris immediately took up the cause of his predecessor/father-in-law. In response to Harris' pleas for more land and more laboratories the leadership of the newly formed LIBA launched in 1925 the first of many fund drives, an ambitious undertaking for such a young organization. Their aim was to raise the money needed to purchase the adjacent 32.5-acre Townsend Jones property. This large tract of land was situated immediately to the north and west of the land that the Bio Lab already leased from the Wawepex Society and had been up for sale for several years. Board chairman Williams led this highly successful inaugural drive which resulted in the purchase of the Jones land in 1926. *(2)*

Counting on the financial support of the Bio Lab's well-to-do neighbors and the moral support of his predecessor, Harris boldly planned to shift the focus of the Bio Lab by augmenting the traditional summer teaching and research activities with a new year-round program in biophysics—a decade-long dream of Charles Davenport. The construction of a biophysics laboratory on the newly acquired Jones property therefore became a top priority. As chance would have it, however, this building would have to wait a little longer.

A new teaching laboratory is urgently needed

The same year the Jones property was acquired an old warehouse on the adjacent Wawepex property which had come to be used by the Bio Lab as a supplementary research and teaching laboratory was destroyed by fire. This whaling era relic had been supplied with "neither gas nor electricity" and according to LIBA secretary Davenport had "little to recommend itself save running water and a good location" at the water's edge on the east side of Bungtown Road. Nevertheless, the facilities it had provided, rudimentary though they had been, were essential for the summer program. As a stopgap measure makeshift research laboratories were set up in Wawepex Building (the old warehouse near Jones Laboratory that had been turned into a lecture hall), but a proper teaching laboratory would have to be built soon. Funds for the construction needed to be raised quickly and an architect had to be found.

Arthur Page commissions Henry Saylor to design some laboratories, mostly Colonial

Even though the intensive campaign to secure the monies needed to purchase the Townsend Jones land had just been completed, the funds to replace the burned building were also successfully raised, a reflection of the growing financial influence of LIBA. The task of finding an architect to design an improved replacement facility fell to LIBA board member and treasurer Arthur Page, who by happy coincidence had excellent connections in the architectural profession. Page's father, Walter H. Page, had founded Doubleday, Page & Company which published books related to art and architecture, and Arthur had worked for the family company early on in his own career as businessman and publisher. As LIBA treasurer and later president (starting in 1927) Arthur Page oversaw the construction of not one but three new laboratory facilities for the Bio Lab, all of which were designed by a former colleague of his at Doubleday, Page & Company, Henry Hodgman Saylor (1880–1967).

Saylor was employed at the Doubleday company as an architectural editor of both books and magazines, including *Country Life in America*. Today, however, Saylor is more widely known for his own books, especially *Bungalows,* published in 1910, and his *Dictionary of Architecture,* first published in 1952 and still available in a paperback edition (see Pictorial

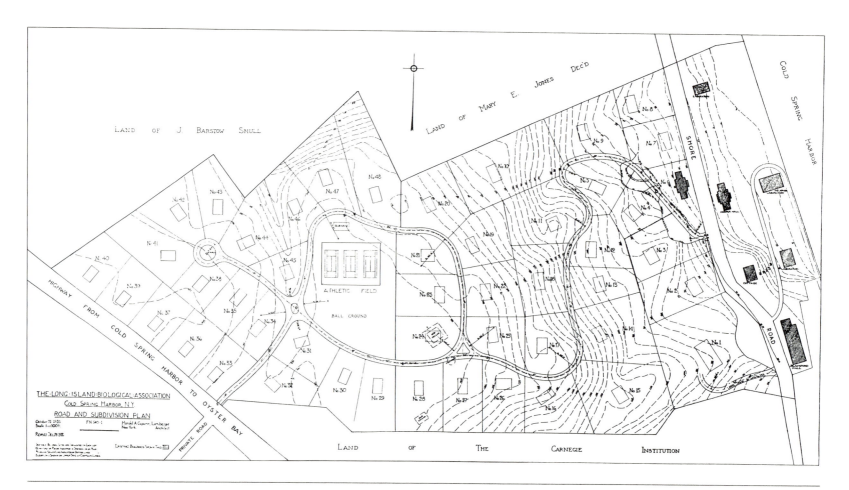

(2) Subdivision plan of Townsend Jones property purchased in 1926.

On this plan published in the *Annual Report of the Biological Laboratory* for 1926 shaded "Existing Buildings" (see enlarged section on the facing page) include the as yet unnamed Williams House on the west (*left*) side of "Shore Road" (Bungtown Road). On the east (*right*) side of the road from north to south (*top to bottom*) were situated the "Laboratory" later named in honor of Charles Davenport, the "J.D. Jones Laboratory" at the harbor's edge, the summer dormitory "Hooper Hall," another "Laboratory" (the former whaling era warehouse later called Wawepex Building), a "Cottage" later called Osterhout Cottage, and the Bio Lab's dining hall "Blackford Hall." The road veering off to the northwest (in the left-hand corner) is Moore's Hill Road, and the "Private Road" intersecting it is Stewart Lane. The subdivision plan shown in this drawing was never executed (see Chapter Four).

Glossary for excerpted entries from this dictionary). Late in his career he became the first full-time editor of the *AIA Journal* (which later evolved into *Architecture* magazine), the widely read official publication of the American Institute of Architects. When he retired from this Washington, D.C., based job he was hailed as America's "dean of architectural editors." *(3,4)*

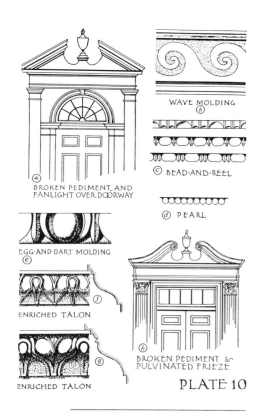

WAVE MOLDING
b

© BEAD·AND·REEL

d PEARL

④ BROKEN PEDIMENT, AND
FANLIGHT OVER DOORWAY

EGG·AND·DART MOLDING
e

ENRICHED TALON
f

ENRICHED TALON
g

⑥ BROKEN PEDIMENT &
PULVINATED FRIEZE

PLATE 10

(3) Drawing from Henry H. Saylor's *Dictionary of Architecture* (1952).

Plate 10 is typical of the illustrations at the back of Saylor's architectural dictionary published in 1952. This was twenty-six years after he designed Nichols Building, whose entrance door enframement bears a certain resemblance to that in detail "h" except that the Nichols Building pediment is not broken and its frieze is not pulvinated (bulging); being flat, it bears the name of the building. (Reprinted by permission from Henry H. Saylor's *Dictionary of Architecture,* copyright 1952 John Wiley & Sons, Inc.)

Although best known for his editing work, Baltimore born and bred Saylor had been trained as an architect at the Massachusetts Institute of Technology. While pursuing his editing career in New York City and raising his family in Huntington, Long Island, he designed a number of single-family houses close to the central business district of Huntington village and also provided renovation plans for older homes in the area, including those of Arthur Page and several of his business partners in the West Hills area south of town. So it was not surprising that Page entrusted Saylor with the design of the three new laboratories needed by the Bio Lab in the mid to late 1920s.

The first building Saylor designed for the Bio Lab, the replacement for the teaching laboratory destroyed by fire, was completed in 1926 and later called Davenport Laboratory in honor of the retired longtime director. (Coincidentally Saylor would design in 1931 a retirement home for Charles Davenport on nearby Laurel Hollow Road.) The remaining two Saylor-designed laboratories went up in rapid succession. Nichols Building, an administration and research facility, was completed in 1928, followed a year later by James Laboratory, which was designed for the newly arrived biophysicist Hugo Fricke (1892–1972), one of Director Harris' most important appointments to the Bio Lab. That same year, 1929, Fricke himself commissioned Saylor to design a small house close to his new laboratory on a small parcel of land that had been part of the Townsend Jones property and which Fricke had purchased.

(4) Detail of the doorway of Nichols Building (see illustrations *7,8,9*).

DAVENPORT LABORATORY

(Now incorporated in
1981 Delbrück and 1987 Page
Laboratories)
Henry H. Saylor
1926

Wearing his architect's hat and using his keen editor's eye, Henry Saylor designed the new teaching laboratory for the Bio Lab along simple, symmetrical Colonial lines so that it would be in keeping with the other buildings along Bungtown Road which were mostly old homes. Looking much like a private house itself, this two-story shingled structure (later named Davenport Laboratory) was erected on the east (harbor-facing) side of the road but turned 90 degrees to it.

The main entrance, flanked by Federal-style sidelights, was on the long side of the building facing south, and the attic gable facing west onto the road featured a half-moon louver. The top floor contained a large botany teaching laboratory, and offices and laboratories for the teaching staff were situated on the floor below. A darkroom and a machinist's shop were located in the basement on the harbor side. *(5,6)*

(5) Davenport Laboratory in the early 1930s.

Built in 1926 as a botany teaching laboratory, Davenport Laboratory was the first scientific building designed for the Bio Lab by the Huntington architect and architectural editor Henry H. Saylor. A simple, symmetrical shingled structure with a gable roof, the building had six-over-six sashes in its windows and an entrance framed by sidelights. These Colonial-style features made the new laboratory building blend in with the old whaling era houses further down the road.

93

(6) Davenport Laboratory today; now part of 1981 Delbrück (*right*) and 1987 Page (*left*) Laboratories.

The structure formerly called Davenport Laboratory occupies the center position in this laboratory complex which is now the headquarters of the plant genetics program at Cold Spring Harbor Laboratory.

NICHOLS BUILDING

Henry H. Saylor
1928

Soon after Davenport Laboratory was completed architect Saylor was commissioned to design a laboratory geared to the research and administrative needs of the year-round program envisioned by Director Harris.

The funds for the laboratory, to be named George Lane Nichols Memorial Building, were donated by banker Acosta Nichols (a nephew of one of the Bio Lab's founders, Franklin Hooper) and his wife in memory

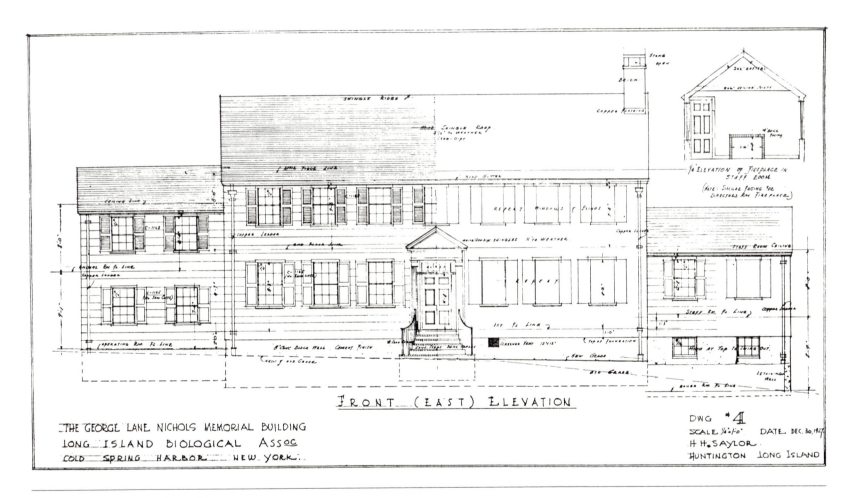

(7) Drawing of Nichols Building.

In the description accompanying this drawing from the *Annual Report of the Biological Laboratory* for 1927 the proposed administration and research building is called "Long Island Colonial" in style. Completed in 1928, Nichols Building was designed by the architect of Davenport Laboratory, Henry H. Saylor, and the two buildings were stylistically similar. Nichols Building, however, had a more elaborate pedimented entrance as shown in illustration 4. Note that paneled shutters were specified on the first floor and louvered ones on the second floor, in authentic Early American style.

(8) Nichols Building in the 1950s.

This photograph was taken only a few years after James Watson (the current director of the Laboratory) spent his first summer at Cold Spring Harbor in 1948 as a researcher in Nichols Building.

of their young son George. Like scores of other neighboring children, George had participated with great zeal in the Children's Nature Study course that had been inaugurated in 1925 by the LIBA Women's Auxiliary chaired by Mrs. Walter Jennings.

Completed in 1928, the two-story shingled Nichols Building featured a handsome pedimented entrance door enframement with a multipane transom window. The building was almost completely symmetrical with recessed wings at both ends,

the single-story wing on the north side containing the Director's Office. The *Annual Report of the Biological Laboratory* for 1927 called it "an example of the Long Island Colonial type of architecture." *(7,8,9)*

Designed for physiology experiments using marine organisms, the new facility contained an Aquarium (located in the main entrance hall on the ground floor), seven private research laboratories, an Animal Room, and an Operating Room.

(9) Nichols Building today.

No longer containing laboratories, this building presently houses many of the administrative offices of Cold Spring Harbor Laboratory.

JAMES LABORATORY
Henry H. Saylor
1929

Erected on the hillside directly behind Nichols Building, the third and last laboratory designed by Henry Saylor at Cold Spring Harbor was the long hoped for biophysics laboratory for Hugo Fricke's work. Named the Walter B. James Laboratory in honor of LIBA's second chairman, it was paid for with monies donated by the recently widowed Mrs. James, a sister of Mrs. Walter Jennings. *(10,11)*

A single-story structure of reinforced concrete in "simple utilitarian" style (to quote the *Annual Report of the Biological Laboratory* for 1929), James Laboratory looked quite unlike Saylor's earlier wooden laboratories with their Colonial embellishments. Its austerity apparently appealed to some, however, for the Bio Lab received at least one request for the architectural plans as a result of a photograph and description of James Laboratory that appeared in the March 1930 issue of *Scientific Monthly*. An accompanying article described how the planned work with X rays (considered fun-

(10) James Laboratory in 1929.

This picture taken at the time of the completion of this laboratory designed by Henry H. Saylor for biophysics work plainly shows its "simple utilitarian" style. Only the door enframement, partially visible behind the cantilevered entrance canopy of reinforced concrete, speaks the "Long Island Colonial" language.

(11) James Laboratory today.
Enlarged by the addition of a
second floor in 1961 and now
flanked by 1971 James Annex
on the left and 1985 Sam-
brook Laboratory on the
right, James Laboratory is currently a locus for tumor virology and cancer research.

(12) Hugo Fricke's biophysics laboratory in James Laboratory.

Featured in this photo is a Wheatstone bridge which was used to measure the electric capacity and resistance of biological materials.

damental for biophysics research) practically designed this building from the inside out:

> The laboratory...has been planned and constructed with special attention to safety, and to the control of temperature, vibration and noise.... It includes an X-ray laboratory equipped with a 10 K.W. 140 K.V., kenotron-rectified X-ray generating unit, installation for pumping X-ray tubes and other kinds of vacuum tubes, apparatus for measuring X rays in the new international r-unit; a laboratory equipped for work with high frequency electric currents; a chemical laboratory equipped with a number of special apparatuses especially designed for a study of the chemical action of X rays, an apparatus for electrometric titration, apparatus for analysis of gases, a spectrophotometer; a mechanics shop with shop equipment and tools for precision work, a carpenter shop and a glassblowers' shop.... (12)

Hugo Fricke investigates the effect of X rays on living cells

The 1929 appointment of biophysicist Fricke signaled the start of a new phase in the intellectual history of the Bio Lab, an era in which research would be conducted on the grounds of the Bio Lab year-round. It was the first step toward the realization of Reginald Harris' dream of being able to financially sponsor a core group of researchers who would remain on the campus after the summer investigators and instructors had left to return to their university posts. Fricke's arrival in Cold Spring Harbor from the Cleveland Clinic was a much heralded event as he was a noted specialist in the field of X-ray and ultraviolet light research. His new well-equipped laboratory became a magnet for physicists interested in biology. Yale-trained Ph.D. Howard J. Curtis became a full-time member of the research group in James Laboratory in 1932, and a year later Kenneth Cole of the College of Physicians & Surgeons of Columbia University began coming in the summer to collaborate with Curtis on the electrical changes that make possible the propagation of signals along nerve fibers.

W.W. Swingle's adrenal cortical hormone relieves
Addison's disease symptoms to universal acclaim

Another researcher working at the Bio Lab in the 1920s and early 1930s was William Salant. Formerly a professor of physiology and pharmacology at the University of Georgia, he was put in charge of a year-round program in experimental pharmacology in Nichols Building. This research focused on the effects of alcohol, bile, strychnine, heavy metals, ergotamine, and ephedrine under both normal and pathological conditions. On the staff during the summer while teaching the Endocrinology course was Wilbur Willis Swingle (W.W. Swingle; 1891–1973), a professor of zoology at Princeton University whose research interest was "internal secretions" (hormones). In 1930 Swingle's work brought instant celebrity to the Bio Lab when it was reported in all the newspapers that a dose of the adrenal cortical hormone that he had isolated at Cold Spring Harbor relieved symptoms of Addison's disease the first time it was administered to a human subject.

The research work in Nichols Building later came under the direction of Edinburgh-trained physiologist Eric Ponder, whose studies focused on permeability, hemolysis (destruction of red blood cells), and the general physiology of red and white blood cells. Ponder first arrived in Cold Spring Harbor in the summer of 1924 as a Carnegie teaching fellow. Prior to joining the staff of the Bio Lab in 1934 he taught at the Washington Square College of New York University. He would eventually succeed Reginald Harris as Bio Lab director.

Harris directs the Bio Lab and at the same time
demonstrates that ovaries contain substances regulating
pregnancy

Director Harris was a man of many talents. First of all he had wooed and won the hand of Jane Davenport, the artistic, bright, and sociable daughter of an impressively dedicated scientist father. Following in his father-in-law's footsteps he soon proved himself to be an excellent administrator and consummate fund-raiser in his own right. His greatest administrative challenge was to locate and hire scientific staff for the three new laboratories and to find the monies to support the research and teaching conducted in them, and this he had done. In the early years of his directorship he even had time to pursue his own research interests, which in the late 1920s led to proof that ovaries contain substances that regulate pregnancy. He was among the first to look for hormones in extracts of the corpus luteum, work that helped lead to the discovery of progesterone by George Corner, a professor at the University of Rochester Medical School who spent summers at the Bio Lab.

Harris and his smallish band of fellow investigators at the Bio Lab were fortunate in that there were other researchers next door at the Carnegie Institution facility to whom they could turn for scientific fellowship and direct collaboration. When Harris' colleague Corner demonstrated that lactation could be induced in spayed female rabbits by an extract from the rabbit anterior pituitary gland, it turned out that the active substance was a hormone that was subsequently isolated, purified, and named prolactin by Oscar Riddle, the Carnegie pigeon man.

Soon there are more researchers at the Bio Lab than students, so more family housing has to be secured

Harris' success in recruiting a year-round staff for the Bio Lab meant that additional staff housing had to be secured as conveniently close to the laboratories as possible. Because he also actively encouraged summertime visits by outside researchers, seasonal accommodations on the Bio Lab grounds had to be expanded as well. At the beginning of the 1930s more laboratory space was becoming available for research because registrations in the summer training courses were declining due to the Depression. Although by the mid-1930s only

(13) Graph of researchers versus students at the Bio Lab as of 1935.

This graph from the Archives of Cold Spring Harbor Laboratory shows the dramatic increase of researchers versus students starting in the late 1920s, the result of a conscious effort by Director Reginald Harris to change the focus of the Bio Lab.

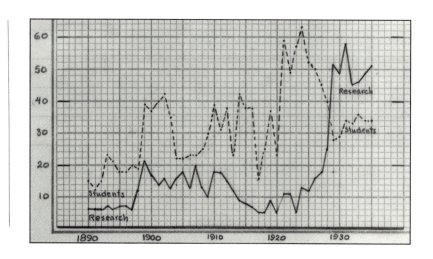

four courses were being offered (typically Surgical Methods, General Physiology, Field Zoology, and Plant Sociology) the educational level of the students was much higher than previously; the majority were graduate students and there were several Ph.D.s enrolled each summer. In addition to the research pursued by the summer visitors formally designated as "investigators," presumably a substantial amount of independent research was carried out by these advanced-level "students."

In 1935 the Bio Lab administration had a graph made for publicity purposes which plotted the number of researchers versus students at the Bio Lab starting in 1890. Both lines generally moved by fits and starts upward to the right, but between the years 1928 and 1929 the ratio of researchers to students did a flip. Prior to 1928 the faster-increasing line on top belonged to the more numerous students, but after 1928 the top line of the graph reflected the more quickly growing number of researchers. For the ever-growing numbers of year-round staff and summer visitors at the Bio Lab various kinds of accommodations were secured, sometimes in rather unique ways. *(13)*

WILLIAMS HOUSE
(Built circa 1835)
Purchased
1926

The on-grounds housing situation was much improved after renovations were made to what the *Annual Reports of the Biological Laboratory* in the late 1920s variously called the "fine old house" and the "charming old house" that was located on the Townsend Jones property purchased in 1926. Originally built in the mid-1830s for the families of workers employed in whaling-related industries, this residence was similar in appearance to Hooper House across the road, originally erected for the same purpose and already used for over thirty years as a summer dormitory. The newly purchased house was later named in honor of Timothy Williams who had led the successful LIBA campaign to purchase the Jones property for the Bio Lab. *(14)*

(14) Williams House in the 1960s.

Built in the mid-1830s as a multiple-family dwelling for workers in the "Jones Industries," both floors of Williams House had typical early nineteenth century door enframements with flanking sidelights. The windows were equally characteristic: twelve-over-twelve on the principal floors of the main block, six-over-six in the wings, and "lie-on-your-belly" horizontal slits in the attic.

(15) Williams House reconstructed.

In 1977 Williams House was
completely rebuilt along the
lines of the original building.
It now contains five two-
bedroom apartments designed
for summer course instructors
and their families but also
suitable for housing year-
round staff on a temporary
basis. Nichols Building is in
the background on the right.

Like Hooper House, Williams House was a long, symmetrical, two-and-a-half-story shingled structure with matching shorter wings on both ends. Both houses were built partly into the hillside and featured porches front and back. Due to the steep change of grade on both sides of Bungtown Road the porches on the sides facing east to the harbor were on the first floors and those facing west were on the second floors. Taller and wider than Hooper House, Williams House featured low, horizontal, "lie-on-your-belly" windows in the attic story and had additional shed-roofed wings abutting the main wings at the ends of the structure.

Williams House must have been in pretty poor condition when LIBA purchased it. The *Annual Report of the Biological Laboratory* for 1928 related that "the roof was reshingled, the chimneys repaired, one ell was rebuilt, the top floor was remodeled and replastered, all of the windows were reputtied, the trim of the windows and doors was repainted, and two bathrooms were installed."

These repairs and renovations were considered a worthwhile investment on two accounts: "these improvements substantially increased our living accommodations...and enlarged our receipts from rental by about fifty per cent."

After central heating was installed in Williams House in 1929 the apartments could be rented year-round to staff members of the neighboring Carnegie facility or to visitors. In time the building also became a winter dining facility. *(15)*

FIREHOUSE
(Built in 1906)
Purchased and relocated to the
Biological Laboratory
1930

In 1930 an unusual opportunity arose for the Bio Lab to acquire more housing both quickly and inexpensively. To raise funds for a new building the Cold Spring Harbor Fire Department planned to auction off the village's 1906 firehouse located on the east shore of the harbor near the west end of Main Street. *(16)*

This oblong, shingled, two-and-a-half-story late Victorian firehouse was plain-looking except for the elaborate facade it presented to the street. On the floor directly above the pair of wide double doors needed for the fire engines were three six-over-six windows in a row and these were flanked on each side by double brackets supporting a heavy arched vergeboard (a highly decorated extension of the eaves of the roof) complete with a carved finial at the top.

At auction Director Harris made the winning bid of $50. He had the firehouse delivered to the Bio Lab by barge and installed at the water's edge just north of the newly erected Davenport Laboratory. This

(16) Firehouse on its original site in Cold Spring Harbor village in the 1920s.

Built on the east side of the harbor in 1906 as Cold Spring Harbor's second firehouse, the building was acquired at auction in 1930 and towed across the harbor to the Bio Lab, where it was installed just north of 1926 Davenport Laboratory and subsequently used as summer housing. Note the elaborate vergeboard decorating the gable end.

was not to be the last trip for this building which is today called simply the Firehouse. Some fifty years later in 1986 it had to be moved to make room for an addition to 1926 Davenport Laboratory. *(17,18)*

The year after the Firehouse made its journey across Cold Spring Harbor the LIBA Women's Auxiliary spent nearly $3300 in refurbishing Bio Lab housing ac-

commodations. According to the 1931 *Annual Report* this large sum, which presumably included many items for the newly acquired Firehouse, covered the purchase of a dozen lampshades and pillows, two dozen blankets, chairs and rugs, and eight dozen new mattresses. Also included in this figure was the cost of renovating two floors of the Firehouse into eleven rentable rooms. *(19)*

(17) Firehouse en route to the Bio Lab in 1930.

A tugboat was used to tow the barge carrying future housing for scientists and their families. The Bio Lab launch is in the foreground and the corner of Jones Laboratory is visible on the left.

(18) Firehouse nearly at the end of its voyage.

The barge is docked at the Bio Lab.

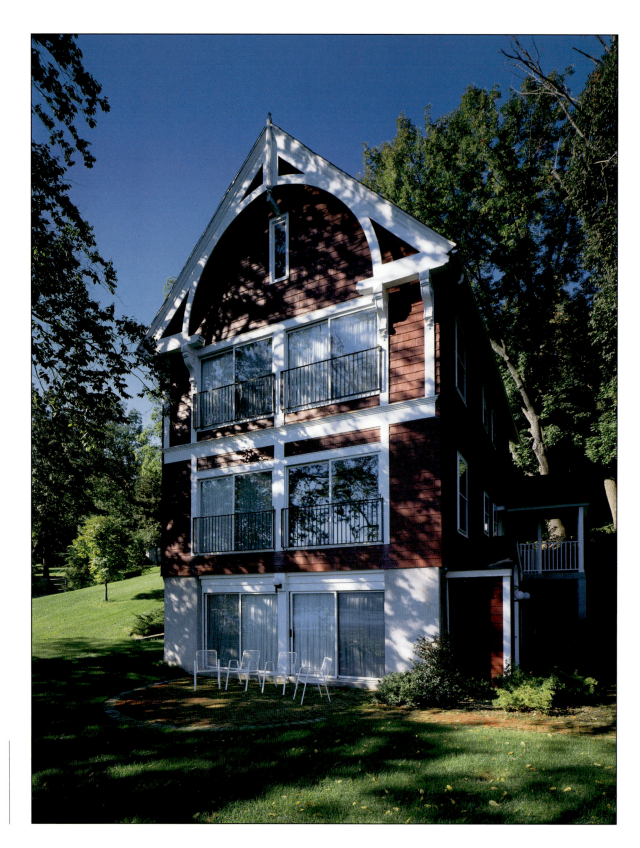

(19) Firehouse today.

Completely renovated in 1972, the Firehouse continues to provide housing for scientists at the Laboratory.

COLE AND UREY COTTAGES
1934

In 1934 two, small, story-and-a-half cottages for summer researchers were completed by a local work crew. These men had lost their regular jobs due to the Depression and had found work at the Bio Lab performing necessary but often deferred maintenance jobs such as painting, reshingling, and general repairs. The two cottages cost less than $1500 apiece to build and were later named after two scientists on the LIBA board who were frequent visitors to Cold Spring Harbor: William H. Cole of Rutgers University and Harold Urey, the discoverer of "heavy water" (which has a higher proportion of deuterium than natural water).

Cole Cottage was subsequently enhanced by the addition of a single-story wing and eventually heated. To this day it continues to provide valuable on-site housing for scientific staff. In the course of a 1978 facelift its entrance was moved to a more satisfactory location and improved by the addition of a front stoop and door modeled after those of historic Raynham Hall in nearby Oyster Bay. *(20,21)*

Meanwhile Urey Cottage grew like "Topsy" to accommodate larger and larger staff families. In the course of a substantial enlargement in 1983 the building gained both a newly fashioned symmetrical

(20) Cole Cottage in the early 1970s.

Built in 1934 as summer housing, over the years Cole Cottage grew larger and gained heat.

(21) Cole Cottage today.

In 1978 the entrance door was relocated and a front door and stoop patterned after those of historic Raynham Hall in nearby Oyster Bay were installed.

exterior, articulated into wings of diminishing size at both ends, and a front porch modeled after that on circa 1835 Hooper House. The style of enlarged and remodeled Urey Cottage could be called neo-"Long Island Colonial," in other words the latest version of the "Long Island Colonial" style that characterizes the old whaling era houses along Bungtown Road and that in the late 1920s architect Henry Saylor went to such great pains to imitate in his laboratory buildings. *(22,23,24)*

(22) Urey Cottage in the 1950s.

Like Cole Cottage, Urey Cottage was built in 1934 as summer housing but it gradually became a full-size house.

(23) Urey Cottage in the early 1970s.

A wing added to the south (*left*) side nearly doubled the size of the cottage.

(24) Urey Cottage today.

Enlarged in 1983 to become the new home of the growing Publications Department of the Laboratory (now Cold Spring Harbor Laboratory Press), the cottage is now symmetrical, in perfect shingled neo-"Long Island Colonial" style.

*Director Harris leaves his mark—the Cold Spring
Harbor Symposia on Quantitative Biology*

As a result of Harris' efforts the Bio Lab entered the Depression years physically
well-endowed in terms of laboratories (three brand-new ones in as many years) and staff
housing. He had also been highly successful in executing his mandate to attract and support a
nucleus of first-rate scientists at the Bio Lab. Making ends meet financially, however, would
soon prove to be an even greater challenge, but one that Director Harris would also meet
head-on in an innovative way that would generate income even to this day. Harris' greatest
claim to fame in the history of Cold Spring Harbor Laboratory and in fact in the history of
modern biology was the inauguration of the world-famous Cold Spring Harbor Symposia on
Quantitative Biology in 1933. He outlined his plan for this new kind of biology meeting in a
December 1932 letter to Bio Lab trustee and noted cell physiologist W.J.V. Osterhout:

> The Laboratory wishes to inaugurate this summer what we believe will be one of
> the most important policies it has adopted and one which we feel will be of the greatest
> importance as time goes on.... The plan is to get together at the Laboratory each summer a
> group of men composed of the leaders in some one branch of quantitative biology.... We
> are planning to have each man give two lectures upon his branch of the problem, the hope
> being that these lectures can be published together in monographic form, a thing which in
> itself would be of great value.... It is our idea to provide for this group comfortable
> quarters without expense for either lodging or food.

*The Symposia bring together the world's best
minds in biology's varied disciplines*

The scheme of the Symposia was, and still is, for biologists in related disciplines—
biochemists, biophysicists, geneticists, microbiologists, and physiologists—to come together
each year for a cross-fertilization of ideas on a single topic. Originally the Symposium lasted
for six weeks each summer; now its length is just one week (in large part due to jet travel).
This is still an extraordinarily long time in today's fast-paced world for a group of scientists,
including international experts, to gather to discuss just one aspect of modern biology. Yet
sequestered along the western shore of Cold Spring Harbor, discuss they do, and with
history-making results each year. The papers presented at the meeting are shortly thereafter
incorporated in the next volume in the ongoing series of maroon-colored books found in
every science library in the world, the *Cold Spring Harbor Symposia on Quantitative Biology*.
Harris' first Symposium, held in Blackford Hall in early June of 1933, focused on the subject
of Surface Phenomena. The next two that followed in subsequent years were on Aspects of
Growth and on Photochemical Reactions. Tragically it was while laboring early in 1936 to

bring out the third annual *Symposia* volume in a timely fashion that Director Harris contracted pneumonia and died at the young age of 38. *(25,26)*

The Symposium, Harris's greatest legacy, typified all of the man's efforts in its labor intensiveness. The effort that went into organizing an international meeting of this scope each year was enormous, particularly for such a small institution as the Bio Lab. There were far too many Symposium participants (upward of several dozen) for all of them to be housed on the Bio Lab grounds, so a number of the scientists had to room in the homes of neighboring LIBA members. Louise Lusk Platt (a granddaughter of Louis C. Tiffany) recalls her parents (the Graham Lusks) "taking in at our old house, The Briars, a great many professors from France, Germany, and England.... The Henry de Forests, at Nethermuir, did likewise." This relationship between meeting participants and the neighbors evolved into what is now a late spring rite at the Laboratory. On the Sunday evening midway through the Symposium (which each year starts on the Wednesday after Memorial Day and runs through the following Wednesday) all of the speakers are invited to dinner parties in the neighborhood.

In the 1930s visiting scientists were also invited by neighbors for private garden tours arranged by Henry Hicks, owner of Hicks Nursery in nearby Jericho, Long Island. Their hosts included LIBA members such as William R. Coe, whose Planting Fields estate in Oyster Bay is now a public arboretum; Henry W. de Forest, president of the New York Botanic Garden and owner of the Nethermuir estate which bordered the Bio Lab at its north end (both the Coe and de Forest estates were landscaped by Olmsted Brothers; the Olmsted planting plan for one of the de Forest houses is shown in illustration *9* in Chapter Four); and Louis C. Tiffany, whose Laurelton Hall north of the de Forest estate was a showplace for the exotic and unusual in domestic and landscape architecture (see photographs in illustration *1*.)

Each year Harris passes the hat

Besides sharing their hospitality with the visiting scientists, the members of LIBA were the single most visible and important source of financial support for the operations of the Bio Lab. The level of research at the Bio Lab was directly proportional to the funds for equipment and salaries that could be raised from the community and private foundations. (These were the days before big government spending for basic science.) In the early 1930s success bred success. The first annual Symposium was so well-received by the biology community that Harris won important backing from the Rockefeller Foundation. They agreed not only to lend a hand in the future support of the Symposia, but also to sponsor collaborative research work by selected participants to be performed at Cold Spring Harbor in the summer weeks following the conclusion of the meeting. A higher level of research activity would hopefully further enhance the Bio Lab's scientific reputation.

Harris' success in steering the Bio Lab along these new paths was largely attributable to his flair for fund-raising, which he began right at home. He did a fine job of convincing the LIBA neighbors to support the Bio Lab, at least until income tax increases and the Depression that followed took their tolls. Prior to the income tax increases individual yearly contributions were measured in hundreds, if not thousands, of dollars, the first large checks having been received in conjunction with the fund drive to buy the Townsend Jones property in 1926. In February of that year LIBA secretary Charles Davenport received the following communique from board member William K. Vanderbilt, the amateur naturalist and yachtsman who made his summer home at Centerport, Long Island, in an elegant Spanish-style mansion called Eagle's Nest (now an historic house museum):

> Your letter of January 27th has just reached me, in which you state that Mr. Schiff has offered to increase his subscription to $10,000 provided the other three principal donors, including Mr. Matheson, Mr. Field and myself, would do likewise.
>
> I should be pleased to increase my subscription to $10,000 and will forward my check to Mr. A.W. Page, c/o Doubleday, Page & Company, Garden City, who I understand is the Treasurer of the Association, whenever you are ready for closing of the title, and in the meantime, as I am going south for a cruise on ARA, I have notified my attorneys, Anderson & Anderson, to be good enough to communicate with you so that in case there is no hitches [*sic*] as far as titles are concerned, my check will be forthcoming in due time through their firm at your request.
>
> Trusting to have the pleasure of seeing you next Summer, I remain
>
> Yours very truly,
> W.K. Vanderbilt

After this fund drive was successfully completed it was back to business as usual in terms of annual financial contributions. On July 21, 1927, Harris received a letter from financier J.P. Morgan that was representative of a vital class of correspondence that crossed his desk. Morgan stated that he was "willing to contribute $500 annually, for three years, to the Long Island Biological Association. I enclose cheque for $500...." Another typical letter, dated September 12, 1930, was received from the secretary of Louis C. Tiffany: "In reply to your appeal of August 26th, Mr. Tiffany is pleased to send a special contribution of One Thousand Dollars ($1,000), as an expression of his interest in the work of the Biological Laboratory, which you have so interestingly described."

Besides solicitation letters, another method of fund-raising was the gala benefit party. Such parties not only raised much needed funds, but they also could garner extensive newspaper publicity for the Bio Lab's cause. On a memorable evening in July of 1932 Bio Lab vice president and treasurer Marshall Field and his wife Audrey hosted a much heralded dinner dance on the lawn of their summer house in Lloyd Harbor for the benefit of the Bio Lab. The following advance publicity ran in the *Huntington, New York Bulletin* on July 15.

> Mr. and Mrs. Marshall Field, who are passing the summer at Caumsett, their estate at Lloyd's Neck, will give an unique charity entertainment this Friday night in their gardens and in their home. The 800 invitations, most of which have been accepted, have

BROOKLYN DAILY EAGLE

NOVEL, COMICS
SPORTS

NEW YORK CITY, WEDNESDAY, JULY 26, 1933

M 2

LONG ISLAND the MAGNIFICENT

L. I. Biologists Study X-Ray Effects

Gather at Cold Spring Harbor and Hope to Establish Foundation for Clearing House of Scientific Information

This is the second in a series of articles on various phases of the colorful and diversified Summer activities and incidents of Long Island life as seen by Harvey Douglass and an Eagle staff photographer in a recent tour.

By HARVEY DOUGLASS

The beautiful North Shore of Long Island, with its rolling terrain, magnificent country estates of the financially mighty, its beautiful harbor and inland lakes and streams, particularly that section centering about aristocratic Glen Cove and Oyster Bay, is not given over entirely to the mere luxury of living.

At Cold Spring Harbor, just west of Huntington, is situated one of the most important institutional centers devoted to scientific lines in the entire East, if not in the United States, and here are now gathered a group of distinguished scientists, both men and women, from the country's leading institutions and from foreign countries, who are giving their entire Summer to their investigations and by which co-ordinated results of wide benefit to humanity at large are the objective. A number of privileged students are working with these experts, bringing the colony of scientific workers to upward of 100 individuals.

This conclave is being conducted under the auspices of the Long Island Biological Association, only one of the several scientific organizations having their laboratories at Cold Spring Harbor, the others—to constitute the topic of future articles in this series—being the Department of Genetics of the Carnegie Institution of Washington; the Eugenics Records Office, a branch of the Carnegie Institution, established through the liberality of Mrs. E. H. Harriman and the largest of the New York State fish hatcheries, all of which have been located here because of the unusual combination of fresh and salt water, permitting the intensive study of all sorts of marine life as well as of a large variety of plant and animal life.

Throughout the month of July the scientists have been supplementing their hours of research with lectures and a symposia on the technically highly important subject of the "Potential Difference at Interfaces and Its Bearing Upon Biological Phenomena," a topic too in-

NOTED SCIENTISTS IN SESSION AT COLD SPRING HARBOR

Above, group of eminent chemists, physicists and biologists conducting research and symposia under auspices of Long Island Biological Association; below, left, Dr. Reginald G. Harris, director of the biological laboratory, and right, memorial erected to Prof. Franklin W. Hooper.

---

(25) Newspaper account of the first Cold Spring Harbor Symposium on Quantitative Biology.

Director Reginald Harris, founder of the Symposia, is shown at the left in this news clipping from the *Brooklyn Daily Eagle* dated July 26, 1933. (Note Hooper House dormitory at the bottom.)

Those Taking Part in Symposia and in Discussions

Abramson, Harold A., Department of Biochemistry, College of Physicians and Surgeons, Columbia University

Barrón, E. S. Guzmán, Assistant Professor of Medicine, University of Chicago

Bates, Robert, Investigator, Station for Experimental Evolution, Carnegie Institution of Washington

Blinks, L. R., Associate, Division of General Physiology, Rockefeller Institute

Briggs, D. R., Investigator in Chemistry, The Otho S. A. Sprague Memorial Institute, University of Chicago

Brownscombe, E. R., Chemist, Biological Laboratory, Cold Spring Harbor

Chambers, Robert, Research Professor of Biology, Washington Square College, New York University

Chen, T.-T., Graduate Student in Physiological Chemistry, (Received Doctorate 1933), Johns Hopkins Medical School

Climenko, Robert, Assistant Professor of Physiology, University of Alberta

Cohen, Barnett, Associate Professor of Physiological Chemistry, Johns Hopkins Medical School

Cole, K. S., Assistant Professor of Physiology, College of Physicians and Surgeons, Columbia University

Curtis, H. J., Physicist, Biological Laboratory, Cold Spring Harbor

Fricke, Hugo, In Charge of Dr. Walter B. James Laboratory for Biophysics, Biological Laboratory, Cold Spring Harbor

Gasser, Herbert S., Professor of Physiology, Cornell University Medical School

Harris, R. G., Director, Biological Laboratory, Cold Spring Harbor

Hill, A. V., Foulerton Professor of the Royal Society; Department of Physiology, University of London

Irwin, Marian, Associate, Division of General Physiology, Rockefeller Institute

Kornhauser, S. I., Professor of Anatomy and Embryology, University of Louisville Medical School

MacInnes, Duncan, Associate Member, Division of Physiological Chemistry, Rockefeller Institute

Michaelis, L., Member, Division of Physiological Chemistry, Rockefeller Institute

Mudd, Emily B. H., School of Medicine, University of Pennsylvania

Mudd, Stuart, Professor of Bacteriology, School of Medicine, University of Pennsylvania

Müller, Hans, Assistant Professor of Physics, Massachusetts Institute of Technology

Osterhout, W. J. V., Member, Division of General Physiology, Rockefeller Institute

Ponder, Eric, Professor of General Physiology, Washington Square College, New York University

Reiner, L., Instructor of Bacteriology, New York University

Riddle, Oscar, Investigator, Station for Experimental Evolution, Carnegie Institution of Washington

Shedlovsky, T., Associate, Division of Physiological Chemistry, Rockefeller Institute

Stiehler, Robert D., Graduate Student in Chemistry, (Received Doctorate 1933), Johns Hopkins University

Svedberg, Theodor, Professor of Physical Chemistry, University of Upsala, Sweden

Van Slyke, D. D., Member, Division of Physiological Chemistry, Rockefeller Institute

---

(26) The roster of Symposium participants as it appeared in that year's *Symposia* volume.

During the month of June 1933, scientists from various disciplines came together at Cold Spring Harbor for the first time to discuss a subject of mutual interest to them all; the topic of that first Symposium was Surface Phenomena. Except for a three-year hiatus during World War II, the Symposia have continued up to the present day, with the proceedings of each being published as the ongoing *Cold Spring Harbor Symposia on Quantitative Biology.* (The asterisks denote papers received and discussed but not presented in person.)

been sent out to a "Dutch Treat Dinner" for the benefit of the Long Island Biological Association. Society folk who have subscribed to the dinner are making up their own tables which are to be arranged on the terrace overlooking the harbor and surrounded by the famous Field gardens, blooms from which have taken many awards in the flower shows both on Long Island and in New York this year.

Besides dancing the entertainment after dinner will include a program of varied features. The former Irene Castle will be there to dance in some original numbers; Vincent Astor will have charge of the china breaking concession; Mrs. Joseph E. Davis of Heydey House, Roslyn, will be in charge of the freak show in which prominent members of society will take part and Neysa McMein, the artist who has her summer house at Sands Point, will conduct a program of living pictures which she will stage with recognizable North Shore characters for subjects.

Society columnist Cholly Knickerbocker wrote up the affair for the *New York Times*, noting that "more than 600 guests were present and dinner parties were arranged by about 100 friends of Mr. and Mrs. Field.... Great flood lights and streamers of vari-colored electric lights arranged in festoon fashion, lighted the lawns and terraces of Caumsett, where the dinner was served by Sherry's...." The proceeds from the dinner tickets (priced at $5 apiece) and the afterdinner attractions (for which the guests paid extra) amounted to thousands of dollars for the Bio Lab. *(27)*

As the 1930s wore on, however, money became tighter and tighter, and Director Harris began receiving communications of a different sort. The example below was penned by one of the Laboratory's most prominent and wealthy neighbors:

> ...While my budget for charities and contributions has already been largely exceeded, in view of what you write I am sending you herewith my check for $100. I wish to say to you quite frankly, that unless the business situation materially improves, I shall make no contribution next year.

Harris sometimes refused to take "no" for an answer, and being a passionate advocate of the miracles of modern endocrinology responded to one rejection letter in the following somewhat rhetorical manner:

> Is it more important to have clothing on the back or food in the stomach of any given person than it is to have adrenal cortical hormone in the blood?

One typical letter, however, was unusually prophetic when it came to the question of the future funding of science.

> I am glad to be able to send the customary check toward the work of the Association, but feel that in sending it I ought to say it will probably be impossible for me to continue the contribution after this year. The reason for this is that I believe the new income taxes to be imposed in the coming year will, when added to the high rates already in effect, completely absorb that portion of my income which in former years I have devoted to

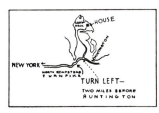

AUDREY FIELD

NEYSA McMEIN

ANNE ALEXANDRE

INVITE YOU TO A DUTCH TREAT DINNER
AND CIRCUS PARTY
(NOT FANCY DRESS)
ON SATURDAY, JULY SIXTEENTH
AT EIGHT O'CLOCK
AT THE MARSHALL FIELD'S

HUNTINGTON
LONG ISLAND

DINNER TICKETS - - - FIVE DOLLARS
THE PROCEEDS GO TO
THE LONG ISLAND BIOLOGICAL ASSOCIATION

FOR TABLE RESERVATIONS TELEPHONE
NEYSA McMEIN'S STUDIO
ENDICOTT 2•9525
MRS. MARSHALL FIELD
HUNTINGTON 1203

(27) Invitation to a party held in July 1932 to raise funds for the Long Island Biological Association.

This benefit dinner dance held at Caumsett, the Lloyd Harbor home of Marshall Field III (grandson of the department store magnate), featured celebrity-run carnival-type concessions for which the guests paid extra.

contributions to worthy causes. However, in the socialized state that seems to lie just ahead of us, such institutions as yours will, I suppose, be no longer dependent on contributions from the private citizen.

As predicted in the letter the burden of support for private institutions, especially scientific laboratories, did in fact shift from private individuals and foundations to a greater responsibility being assumed by the federal government, but this did not occur until after World War II.

The Bio Lab celebrates turning 50

As the decade of the 1930s drew to a close it was not clear whether the Cold Spring Harbor community was aware of the increasing national and international importance of the Biological Laboratory that had grown and matured in its midst. This matter was addressed in one of the articles in the commemorative booklet of the Bio Lab's Semi-Centennial, celebrated in late June 1940. The article pointed out that the best way to illustrate the influence of the Biological Laboratory in the scientific world of that day would be to construct "a map showing where, considering North America alone, our research workers and students have come from during the last five years.... One could make another map, showing the places to which our Symposium volumes are distributed, but this would have to be a map of the world with every important center of science marked upon it." The article went on to quote a speech given several years earlier by LIBA chairman Arthur Page:

> I have a feeling that many of our neighbors do not know how important an institution has grown up unobtrusively on the harbor.... Perhaps this is because our ratio of biology to buildings is higher than usual and buildings are easier to see than biology—perhaps only because, like the prophets, biologists may lack honor in their own country.
>
> But of this I am certain—if the neighbors who pass the Laboratory to sail and golf or go to Dr. Bleecker's church should go to Scotland and there be shown an institution which did just what is done at Cold Spring Harbor—and there is none in Scotland—they would come home and say how advanced the Scots are in science....
>
> And this Laboratory is peculiarly a community enterprise, for with the exception of money granted by the Rockefeller Foundation and the Carnegie Corporation during the Depression—for which grants we are extremely grateful, both because of the money, which was essential, and for the endorsement, which was most gratifying—the Laboratory has been supported by the people of the neighborhood who have had an interest in this method of furthering the progress of mankind.

The Depression continues, war threatens, and the
Harris-Ponder era of physiological research ends

Upon the premature death of Reginald Harris in 1936 staff scientist and physiologist Eric Ponder was made director of the Bio Lab. The summers at Cold Spring Harbor continued to be intensely flavored by the Symposia, now organized by Ponder, for each year ongoing Rockefeller Foundation funding allowed a number of the participants to remain in residence for the whole season. After the Symposium on Excitation Phenomena in 1936 F.O. Schmidt and Richard Beck from Washington University stayed on to collaborate with

J.Z. Young, the Oxford zoologist who had just discovered the large nerve fibers of the giant squid. In 1937, however, it was not the Symposium on Internal Secretions but Ponder himself who brought the English physicist W.T. Astbury, the English physiologist Hugh Davson, and the German-born biochemist Hans Neurath to the Bio Lab that summer. The 1938 Symposium was devoted to Protein Chemistry, and Neurath and Davson were both again in residence, as was the English chemist J.F. Danielli. Davson and Danielli later coauthored a very influential book entitled *The Permeability of Natural Membranes* (1943). By the time the Symposium on Biological Oxidations was held in 1939 war was about to break out in Europe, and Fritz Lipman, who had arrived from Copenhagen for the meeting, did not return home. European-born Carl Cori, who came from St. Louis with his wife Gerty, later remembered being assured by a local resident whom he met in a bar across the harbor that peace would hold until the end of the summer. His informant, a driver for J.P. Morgan, was certain that Mr. Morgan would not let the war start until he had returned home to his estate in Glen Cove, Long Island.

No Europeans could come for the 1940 Symposium on Permeability and the Nature of Cell Membranes, which turned out to be the last Symposium organized by Ponder. The continuing Depression and the increasing likelihood that the United States would enter the war brought local community support effectively to an end. There were no funds to pay biophysicist Fricke's salary and he departed for war-related work. As the year drew to a close Ponder was the only scientist left at the Bio Lab. He had no choice but to resign, which he did to the LIBA board at its November 1940 meeting in New York City.

CHAPTER FOUR

The Carnegie Department of Genetics and the Biological Laboratory share resources and later join forces to become the Cold Spring Harbor Laboratory of Quantitative Biology

1940~1970

Eric Ponder's resignation from the directorship of the Biological Laboratory became effective in 1941 and that same year Albert Blakeslee retired from his post as director of the Carnegie Department of Genetics. Both directorships were then assumed by Milislav Demerec, the plant geneticist turned *Drosophila* specialist who had worked at Cold Spring Harbor since 1923. The Demerec era (1941–1960) would be one of closer cooperation than ever before between the two scientific institutions. As World War II played out on the battlefields of Europe the Carnegie genetics program continued to expand while the Bio Lab, lacking resources, was forced to retrench to its earlier summer-only program of courses and the Symposia. Then the unexpected happened. Cold Spring Harbor found itself cast into the new role of providing a scientific refuge for some of Europe's finest experimentalists fleeing totalitarian tyrannies in their homelands. By the end of the war these scientists had forged a brand-new discipline in the biological sciences— "molecular biology." The process of this evolution and the demands that the new science made on its practitioners forever changed the flavor and pace of scientific life at Bungtown.

Chemically Defining the Gene

By 1941 the key questions facing geneticists were (1) the chemical identity of the gene, (2) the mechanism by which genes are duplicated when cells grow and divide, and (3) how genes control the properties of the cell in which they reside. To answer these questions a number of genetically oriented scientists began to question whether any decisive answer could in fact be obtained using the already familiar model systems such as *Drosophila*, maize (corn), and the mouse. Had not the time come to study much more intensively the genes of simpler single-cell microorganisms such as bacteria and fungi? At the vanguard of the microorganism-oriented attack were the German theoretical physicist Max Delbrück and the medically trained Italian Salvador Luria, both of whom had fled the totalitarian regimes then engulfing Europe. The microorganism they focused on, the bacterium *Escherichia coli* (*E. coli*), had multifold advantages. It was easy to grow using defined nutrients, was nonpathogenic, and was host to a variety of viruses (bacteriophages), which themselves might be viewed as virtually naked genes. Until Luria and Delbrück's early 1940s experiments bacteria were thought to be genetically unlike higher organisms and possibly not to have conventional genes. Their experiments decisively brought bacteria into the genetics arena by showing that they have real genes that can spontaneously mutate. Soon after, in 1946, Joshua Lederberg and Edward Tatum at Yale showed that bacteria have sex and that genetic recombination can occur between different bacterial strains.

E. coli was not the microorganism that provided the chemical identity of the gene, however. The crucial answer came from work on pneumonia bacilli. In 1928 the English bacteriologist Fred Griffith made the very unexpected observation that nonpathogenic cells could be hereditarily changed into pathogenic cells by a factor coming from heat-killed pathogenic cells. The search for the chemical identity of the hereditary factor lasted until 1944 when Maclyn McCarty, working in Oswald Avery's laboratory at the Rockefeller Institute, found that it was DNA (deoxyribonucleic acid). DNA had long been known as a major constituent of chromosomes in both plant and animal cells, but whether one or more of the protein components of chromosomes carried genetic information was unknown.

Earlier, DNA had also been found to be a major component of the bacteriophages studied by Luria and Delbrück, and it was natural to presume that it was also the genetic molecule of these bacteriophages. Proof for this conjecture came in 1952 when Alfred Hershey, working at Cold Spring Harbor, showed that following the attachment of a bacteriophage to a bacterium only the DNA component enters the bacterium. This ruled out any genetic role for the protein that remained outside the bacterial cell and so played no role in the transmission of hereditary characteristics.

DNA was still very mysterious when it was found to be the gene. It was known to be a long, thin polynucleotide built up from hundreds if not thousands of nucleotide building blocks. Four different nucleotides exist containing, respectively, the bases adenine, guanine, cytosine, and thymine. Initially the exact chemical bonds that held the nucleotides together were unknown. Alexander Todd's laboratory at Cambridge solved the linkage problem by 1951, and only two years later James Watson and Francis Crick, also working at Cambridge, proposed that DNA was a double helix in which the two polynucleotide chains are held together by weak chemical bonds (hydrogen bonds between adenine and thymine and between guanine and cytosine). The A-T and G-C base-pairing rules generate complementary base sequences on the two chains. The sequence of one chain automatically dictates the sequence on the complementary chain. Immediately obvious was the proposal that the long-sought mechanism of gene duplication involves separation of the two strands of the double helix. The resulting single polynucleotide chains could then serve as molds (templates) on which chains with complementary sequences could assemble to form daughter double helices. Proof for this hypothesis came only five years later when in 1958 Matthew Meselson and Franklin Stahl at the California Institute of Technology experimentally demonstrated that the strands of parental double helices do in fact separate during gene duplication. Equally important was the 1956 discovery by Arthur Kornberg in St. Louis of an enzyme (DNA polymerase) that would replicate DNA in cell-free extracts. With this discovery the study of DNA became the province of the biochemist as well as the geneticist.

The Symposium focuses on genetics for the first time

When Milislav Demerec (1895–1966) became the director of both the Biological Laboratory and the Carnegie Department of Genetics at Cold Spring Harbor in 1941 it came as no surprise that the Bio Lab would need to abandon for the time being any plans for continuing a program of year-round research. This was the only prudent course since it had no endowment with which to guarantee staff salaries and in the wake of the Depression little prospect of one. Demerec might make new appointments to the Department of Genetics, but the role of the Bio Lab, given its sparse resources, would be to concentrate on running the annual Symposium (by then a more manageable two weeks in length rather than the original six) and to sponsor the work of summer researchers on a limited basis.

Demerec broke new ground when he chose a genetics topic, Genes and Chromosomes, for his initial Symposium. This 1941 Symposium proved to be a landmark in the history of genetics for it was the first time that geneticists were brought together with the structural biologists who had just begun to study the nucleic acids DNA and RNA. Among those present was Sewall Wright, the most mathematically astute American geneticist, who had first come to Cold Spring Harbor as a student in 1914. Also in residence was the chemist Wendell Stanley, who later won a Nobel prize for his isolation and crystallization of the tobacco mosaic virus. The concluding talk was given by Hermann J. Muller, who had just returned to the United States after spending ten years in Europe, first in Berlin, then in Moscow, and most recently in Edinburgh where he had arrived in 1939 to attend the International Congress of Genetics.

Following the 1941 Symposium eleven participants stayed on to do experiments on *Drosophila*. A small group of maize geneticists—among them Barbara McClintock from Missouri (later appointed to the Carnegie staff), Harriet Creighton from Wellesley, and Marcus Rhodes from Columbia University—also remained and collectively grew more than 10,000 maize plants for investigation of their chromosomes. That summer also provided the first opportunity for Alfred Mirsky of the Rockefeller Institute to collect trout sperm from the Fish Hatchery to study their DNA components and for the physicist Max Delbrück and the microbiologist Salvador Luria to initiate their joint experiments on bacteria-eating viruses (bacteriophages) in Jones Laboratory.

During the war years only Jones Laboratory would remain available for summer research and it was to there, straight from his Vanderbilt University teaching duties in Nashville, that Max Delbrück would return each summer with his wife Manny to do further experiments on bacteriophages with Salvador Luria, then at Indiana University. Both Delbrück and Luria were citizens of enemy countries and therefore not eligible for military service or secret war-related research so they took advantage of the opportunity afforded by the summer research program at Cold Spring Harbor to continue their own research.

Cold Spring Harbor researchers aid the war effort

Due to the hostilities in Europe there were no Symposia in 1943, 1944, and 1945 so Demerec turned his thoughts to the war effort and how the laboratories at Cold Spring Harbor might be able to help. Just before the United States entered World War II he arranged for the Airborne Instruments Laboratory of Columbia University to take over James and Davenport Laboratories and Urey and Cole Cottages for work on antisubmarine defenses. Nichols Building was put at the disposal of the Chemical Warfare Service of the War Department for Macy Foundation-funded experiments on the fine mists (aerosols) used to dispense poisonous gases into the atmosphere. There Vernon Bryson (later to join the Bio Lab staff) helped to develop the Cold Spring Harbor Aeroliser, a device that subsequently proved useful in administering antibiotics instead of poisons.

The major wartime success at the Carnegie Department of Genetics was the creation of X-ray-induced mutations that increased the yield of the newly discovered drug penicillin. Demerec, working with Eva Sansone and Harold Warmke, produced a genetic variant that yielded twice as much of this life-saving antibiotic as the original fungal strain studied at Oxford by Howard Flory. Salvador Luria remained all year at Cold Spring Harbor between 1944 and 1945 to work on a War Production Board-sponsored project that studied bacterial variants resistant to penicillin and streptomycin. (He and his wife Zella made their home in de Forest Stables.) There were also attempts by Warmke, supported by the Bureau of Plant Industry, to breed a hemp fiber plant with a negligible marijuana content, but this project was eventually abandoned.

Carnegie researchers continue to pursue different
approaches to genetics

In the meantime Demerec had organized the ongoing work of the Carnegie Department of Genetics into four main sections. The maize cytogenetics effort was led by Barbara McClintock, who formally joined the staff in December of 1941, one of Demerec's first appointments as director. The continuing research on the cytogenetics of *Drosophila* was placed in the hands of Berwind Kaufman, who was newly hired from the University of Alabama. Carleton MacDowell persevered in his research on the heredity of cancer in mice. Demerec's own work now centered on the nature of the gene as revealed by the way it mutated to new forms and by the end of the war he had begun experiments on chemical mutagenesis, focusing on several mustard gases used in World War I. As World War II neared its end in 1944 Charlotte Auerbach in Edinburgh found that these gases were powerful mutagens in *Drosophila*.

With the new appointments to the year-round Carnegie staff came the need for additional housing for the scientists, and the around-the-clock pace of science demanded that staff housing be as close to the laboratories as possible. Luckily for science at Cold Spring Harbor several unexpected possibilities presented themselves in the early 1940s.

Mrs. Henry de Forest donates the Sand Spit and
nearby property to the Bio Lab

Long before the first laboratories were built along Bungtown Road members of the de Forest family owned the lands at its north end, including the Sand Spit which jutted out into Cold Spring Harbor from its western shore. Henry Wheeler de Forest (1856–1938) was the second generation of his family to summer at Cold Spring Harbor, and for many years it had been with his express permission that the neighboring scientists enjoyed the use of the bathing beach on the north side of the Sand Spit (see Chapter One). The centerpiece of the de Forest's country estate was their home Nethermuir, a large, classically detailed, mid-nineteenth century farmhouse to which they made several additions.

Trained as a lawyer, Henry de Forest rose to prominence in the worlds of banking, business, and the railroads. He was also a celebrated fund-raiser for medical and conservation causes, helping to found the Columbia-Presbyterian Medical Center in New York City and serving for many years as president of the New York Botanic Garden. Wishing to enhance the grounds of his estate in Cold Spring Harbor de Forest contacted the celebrated landscape architects Olmsted Brothers for the first time in 1906. (This was just after Frederick Law Olmsted, Jr., the founder's son, was consulted regarding the siting of Blackford Hall [see Chapter One].) Over the course of the next twenty years the Olmsted firm prepared more than 250 drawings for the de Forest family, including numerous site plans and planting plans. In addition to extensive gardens, the grounds also contained several major structures besides the main house. One of these, a stables building, was donated to the Bio Lab by Henry de Forest's widow in 1942, four years after her husband's death. The nearby Sand Spit and the parcel of land on which the stables stood were also part of the gift.

DE FOREST STABLES
(Designed by Clinton Mackenzie;
Built in 1914)
Acquired
1942

The carriage house and stables building on the grounds of the de Forest estate was erected in 1914 to the designs of Clinton Mackenzie, a noted Long Island landscape architect. (Mackenzie also designed the gardens and several of the outbuildings at Fort

(1) De Forest Stables.

A concrete structure with vertical boards sheathing the second story, this stables building was erected in 1914 on the Henry W. de Forest estate just north of the Biological Laboratory. The architect, Clinton Mackenzie, also designed the gardens and many of the outbuildings at Fort Hill, the Lloyd Harbor estate of

Bio Lab president William Matheson. The building, together with the nearby Sand Spit, was donated to the Laboratory in 1942 by Henry de Forest's widow and it has been used ever since as year-round housing for scientific staff.

(2) View of de Forest Stables from the summertime communal garden of Cold Spring Harbor Laboratory.

Hill, the Lloyd Harbor estate of Bio Lab president William Matheson.) A U-shaped two-story concrete structure with vertical wooden sheathing on the upper floor, the stables housed both horses and automobiles on the ground floor and contained apartments for estate staff on the floor above. *(1)*

Known today as de Forest Stables, the building has been in continuous use as housing for scientific staff since the time of Mrs. de Forest's gift. In response to changing needs it has undergone numerous minor renovations, especially the reallocation of space by swapping bedrooms among the three apartments. The most recent renovation was the creation of a garden apartment in the southwest corner of the previously uninhabited ground floor. In the description of the acquisition of the stables and surrounding land that appeared in the *Annual Report of the Biological Laboratory* for 1942 it was noted that "a portion of the nine-acre tract has been used as a vegetable garden in the past, and makes a very suitable site for an experimental garden." Today the Laboratory's summertime communal garden is located southeast of de Forest Stables. *(2)*

Another de Forest house, Airslie, is acquired by the Bio Lab

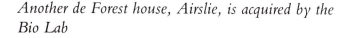

Besides the stables and the de Forests' main residence Nethermuir (demolished in the mid-1940s) the estate incorporated a smaller, older farmhouse that directly overlooked the Sand Spit. This house belonged to Henry de Forest's sister Julia and as documented in an old map had been known as "Airslie" since at least the mid-nineteenth century. Miss de Forest was a talented amateur photographer, dedicated feminist, and effective fund-raiser for the New York Infirmary for Women and Children. Since her principal residence was at her brother's home in Manhattan the house was sometimes loaned to friends and relatives. When Julia de Forest's good friend Louise Wakeman Knox married widower Louis Comfort Tiffany in 1886 the couple spent the first weekend of their honeymoon at Airslie. *(3)*

Nearly sixty years after its use as the Tiffany's honeymoon retreat, Airslie, like the de Forest Stables, became the property of the Bio Lab, but this time through a deed of sale rather than as a gift. Purchased in 1943 for $10,000 together with the seven acres on which it stood, Airslie became the residence of Bio Lab director Milislav Demerec and of all subsequent directors. Long before it came into the possession of the de Forest family, however, this farmhouse next to the beach had a long past rich in associations with another distinguished family in the area who had even closer connections to the Bio Lab.

(3) De Forest family sampler dated 1928 showing Airslie (*upper left*) and Nethermuir (*lower center*), the two homes on the Henry W. de Forest estate north of the Bio Lab.

According to the caption cross-stitched in the center of the sampler it was "made by Julia Noyes de Forest 1928." She was the wife of Henry Wheeler de Forest (1856–1938) and they occupied Nethermuir, the elegant main house, until de Forest's death in 1938. The smaller farmhouse Airslie belonged to his sister Julia. Since 1943 it has served as the home of successive directors at Cold Spring Harbor. Other captions in the sampler include "A garden is a lovesome spot" (*top right*) and the names of the two generations of de Forest families that lived at Nethermuir, "Henry Grant & Julia Mary de Forest 1866" (*bottom left*) and "Henry Wheeler & Julia Noyes de Forest 1901" (*bottom right*).

Airslie was originally built for a member of the Jones family of Cold Spring

The house the Bio Lab purchased for Director Demerec and his successors had been built in 1806 for Major William Jones (1771–1853), a former cavalry officer and a first cousin, once removed, of Bio Lab founder John D. Jones. Unlike most of the local Joneses who were involved in commerce, manufacturing, and whaling at Cold Spring Harbor, Major Jones was active in countywide agricultural affairs and was a founder in 1817 of the Queens County Agricultural Society. According to the Jones family genealogist he was fond of thoroughbred horses and "next to his family, he loved his horse." Local residents, and even an old map (see Prologue, illustration *10*), referred to the half-mile-long stretch of beach on the Sand Spit below his house where he exercised his horses as "Major Jones' Beach." *(4)*

In 1790 Major Jones married Keziah Youngs. She had been raised in the shingled farmhouse (later owned by members of the Roosevelt family) at the northwest corner of

Cove and Cove Neck Roads in nearby Oyster Bay. Here George Washington spent the night of April 23, 1790, on his celebrated post–Revolutionary War tour of Long Island and Keziah Youngs figures in this story. To quote from *"Walls Have Tongues": Oyster Bay Buildings and Their Stories* edited by John MacKay, "Nine generations of the Youngs family owned this house...and its proudest day...was when George Washington 'rested' there on his Long Island tour. The Father of Our Country kissed young Keziah Youngs on the cheek, which, the rest of her life, she 'held sacred to his kiss'."

Sixteen years after his marriage to Keziah Youngs Major Jones built his large farmhouse above the Sand Spit, and in this spacious home the Joneses raised a family of eight and entertained frequently enough (especially politicians such as De Witt Clinton) for their home to gain a reputation as "an open house, where always was to be had good brandy and Holland gin." The house was later occupied by Major Jones' son David William Jones, who bred horses like his father and also wrote about them. *(5)*

(4) Airslie from the Sand Spit.

Airslie was built in 1806 for Major William Jones, a cousin of Bio Lab founder John D. Jones and a renowned breeder of horses. He trained them below his house on the Sand Spit beach, which in those days was known as "Major Jones' Beach."

(5) Airslie in early evening.

According to the family genealogist John Henry Jones (*The Jones Family of Long Island,* 1907) Major Jones' home was "an open house, where always was to be had good brandy and Holland gin."

AIRSLIE
(Built in 1806)
Purchased
1943

The large shingled farmhouse that was built overlooking the Sand Spit for Major Jones and his family was in up-to-the-minute Federal style. Designed by an anonymous carpenter/builder, the house featured a broad, heavy, paneled front door framed by slim, reeded pilasters and sidelights filled with 16 panes each of the thinnest glass. The two-and-a-half-story residence was covered by a Dutch-influenced gambrel roof (two pitches) and had exposed brick chimneys at both ends of the main section.

By 1855 the house belonged to a "G.L. Willard, Esq.," according to a map of that date, and was called Airslie. A sketch of the house on the 1855 map shows that by this time Airslie had gained a long two-story ell extending out from the west (rear) wall of the house. The addition's pitched roof was broken on the south side by a small gable that contained an arched window, and a similar detail had been added on the main facade facing east onto the Sand Spit. Single-story octagonal bay windows had also been added at both ends of the house. These were later raised to two stories, presumably when Stephen Linington and his family lived there in the 1870s and 1880s. He was a brother of the Timothy B. Linington who built the large Victorian residence (today known as Olney House) just down the road from Airslie. *(6,7)*

It was under the later ownership of Henry Wheeler de Forest, with his strong botanical interests, that the Airslie lawn was

(6) Bay windows of Airslie, south facade.

The octagonal bays, a mid-nineteenth century addition (see illustration 7), were extended to the second story in the late nineteenth century.

MAP of LAND
belonging to
G. L. WILLARD, Esq.

OYSTER BAY L.I.

Contains 18 A. 1 R. 14 r.

AIRSLIE

Surveyed June 1855
Magnetic Variation 5° West. Chain 66 ft.
J. J. Shipman Surveyor
Huntington L.I.

Scale 1½ Inches to the Chain

Land of David W. Jones

Land of O. H. Jones

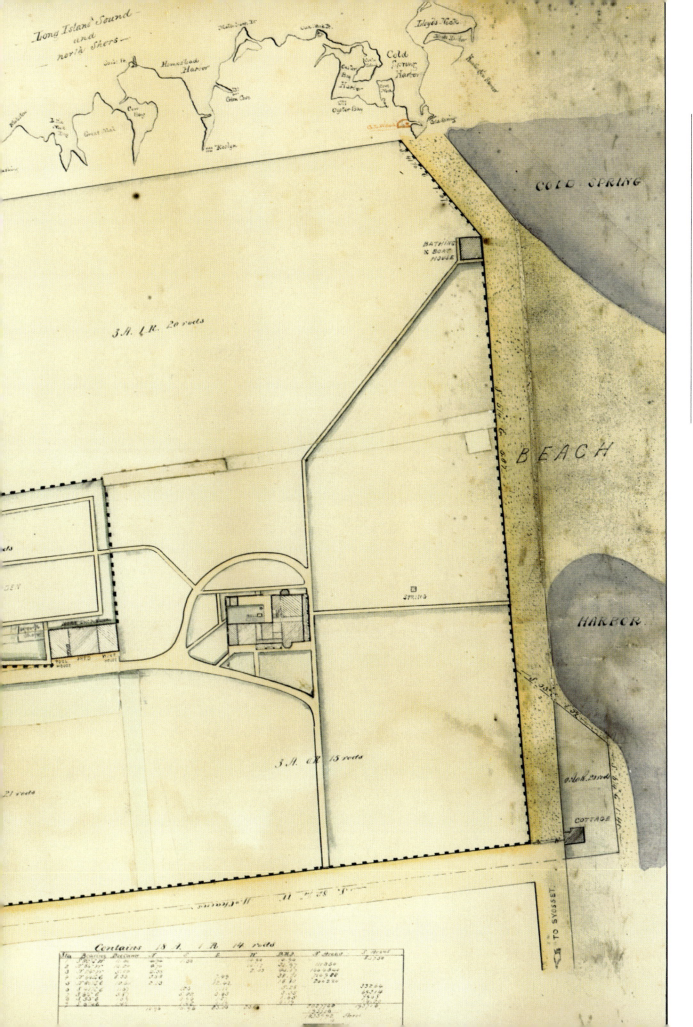

(7) "Map of Land belonging to G.L. Willard, Esq." 1855.

This map of the approximately 18 acres that Willard owned in the vicinity of Airslie features an attractive rendition of the house in its mid-nineteenth century Gothic Revival guise. Note the four-over-four windows (meant to look like casements) trimmed in heavy drip molds on their tops and sides and the mock-Tudor-style porch. Single-story octagonal bay windows together with a cross gable centered on the main facade complete the picture.

(8) Topological plan of the "Henry W. de Forest Estate" drawn in 1906.

This was a baseline plan to show existing conditions prior to commencement of work by the celebrated firm of landscape architects Olmsted Brothers. Nethermuir is on the left, Airslie on the right. In the course of twenty years the project generated over 250 drawings which now reside in the archives of the Frederic Law Olmsted National Historic Site in Brookline, Massachusetts.

Henry W. de Forest Estate
Oyster Bay - Nassau Co. - New York
Topographical Plan
of Grounds near & adjoining Residence

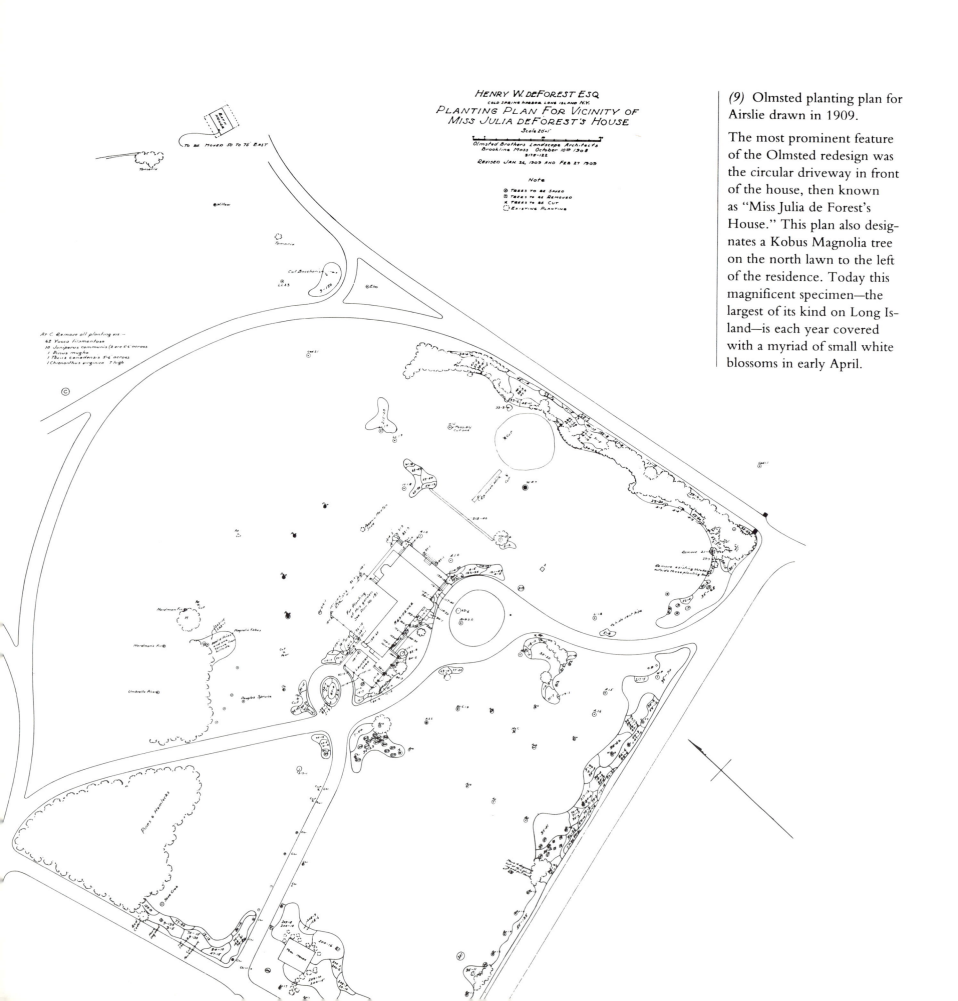

HENRY W. deFOREST ESQ
COLD SPRING HARBOR, LONG ISLAND N.Y.
PLANTING PLAN FOR VICINITY OF
MISS JULIA deFOREST'S HOUSE
Scale 20=1'

Olmsted Brothers Landscape Architects
Brookline Mass October 10th 1908
3175-122
Revised Jan 26, 1909 and Feb 27 1909

Note
Ⓢ Trees to be Saved
Ⓡ Trees to be Removed
✕ Trees to be Cut
◯ Existing Planting

(9) Olmsted planting plan for Airslie drawn in 1909.

The most prominent feature of the Olmsted redesign was the circular driveway in front of the house, then known as "Miss Julia de Forest's House." This plan also designates a Kobus Magnolia tree on the north lawn to the left of the residence. Today this magnificent specimen—the largest of its kind on Long Island—is each year covered with a myriad of small white blossoms in early April.

(10) Airslie in the 1960s.

The circular driveway designed by Olmsted Brothers after the turn of the century is still in place. The various windows tell the building's architectural history, from the twelve-over-twelve sliding sash windows of the Federal period, through the four-over-four sash windows popular in the mid-nineteenth century, to the later Victorian two-over-two windows.

planted with the specimen trees that look so amazing today. Among these are two majestic examples of rare imports from the Orient, an Amur Cork tree and a Kobus Magnolia, which together with a mature American Horse Chestnut comprise a trio of trees on the north side of Airslie that are among the largest specimens of their kind on Long Island. With their huge branches dipping into the ground and growing back out again these old trees have provided a dramatic backdrop for the picnics and other parties for summer visitors that have been held on the lawn of Airslie since the time it became the official residence of the Laboratory director starting with Milislav Demerec in 1943. *(8,9,10,11)*

(11) Airslie today from the apple orchard.

Delbrück starts the Phage Course in 1945

By the end of World War II Director Demerec had through his contacts with Delbrück and Luria been partially seduced away from *Drosophila* to work with *Escherichia coli*. To ensure that the new techniques for studying *E. coli* and its bacteriophages would be available to a larger audience, Demerec invited Delbrück to teach a training course at Cold Spring Harbor in the summer of 1945. This was the start of the now historic course on bacteriophages, which was quickly nicknamed the "Phage Course." Like the earlier Bio Lab courses in the late 1920s and 1930s, it was taught in Davenport Laboratory (today a part of 1981 Delbrück and 1987 Page Laboratories). The Phage Course not only marked the revival of summer teaching at Cold Spring Harbor (and on an even higher plane since most of the students were Ph.D.s), but it also signaled the arrival of a new era in biology, that of "molecular biology"—a synthesis of physics, biochemistry, microbiology, and genetics.

Over the years the Phage Course numbered among its students many who went on to distinguished careers in molecular biology, including Seymour Benzer, Bernard Davis, and Gunther Stent. The Hungarian-born theoretical physicist Leo Szilard, who was involved with Enrico Fermi in building the first atomic pile, attended the course in the summer of 1946, the third year it was offered. Norton Zinder took the course in 1949, just three years before his discovery that bacteriophages could be used to transfer genes from one bacterial strain to another.

Molecular biologists assemble for the early post-war Symposia

Equally important for the emerging field of molecular biology were the immediate post-war Symposia. At the 1946 Symposium on Heredity and Variation in Microorganisms, which attracted future Nobel laureates Jacques Monod, Edward Tatum, André Lwoff, and Joshua Lederberg, the discoveries of sex in both bacteria and bacteriophages were announced for the first time. Participants at the 1947 Nucleic Acids and Nucleoproteins Symposium, including the English chemist John Gulland, the Belgian Jean Brachet, and Viennese-educated Erwin Chargaff, then analyzing DNA at the College of Physicians & Surgeons of Columbia University, were treated to a busman's holiday visit to Brookhaven National Laboratories, an hour's drive east of Cold Spring Harbor. The focus of the 1948 Symposium on Biological Applications of Tracer Elements was on the many important uses of newly available radioactive isotopes. The 1949 Symposium on Amino Acids and Proteins, which attracted among others Fred Sanger, the British biochemist who was working out the structure of insulin, featured a pair of quintessential Bio Lab-style summer delights—square dancing, with Carnegie mouse researcher Carleton MacDowell as caller, and a picnic at Jones Beach.

(12) Ears of maize (Indian corn) similar to those studied by Barbara McClintock.

Through her experiments with maize plants McClintock demonstrated the existence of "jumping genes," a phenomenon now known to be universal in the biological world. (Photographed in Page Laboratory.)

Barbara McClintock announces "jumping genes" at the 1951 Symposium

To an incredulous audience at the 1951 Symposium on Genes and Mutations Barbara McClintock presented evidence that movable genetic switches—"jumping genes"—can turn genes on and off. This finding stemmed from her work on maize which she had begun less than ten years before, right after her appointment to the Carnegie staff. However, because Mendelian genetics describes genes that do not move, the applicability of her findings to other organisms remained unclear for many years. It was not until the 1980 Symposium on Movable Genetic Elements—three decades of Symposia later—that the existence of "jumping genes" was accepted as a general biological phenomenon. McClintock received her Nobel prize not long afterward in 1983. *(12)*

European-born scientists make the Bio Lab their summer home

In the years during and just after the war the large number of prominent European-born biologists and chemists living in New York City provided a valuable resource for Cold Spring Harbor. Demerec not only sought their advice on scientific matters, but also encouraged them to spend summers along Bungtown Road with their families. First to take up summer residence was the German-born ornithologist Ernst Mayr, then at the American Museum of Natural History. Beginning in 1943 he spent ten summers living in either Williams House or Hooper House. The Russian geneticist and evolutionist Theodosius Dobzhansky was in residence between 1944 and 1946. In the latter year the Viennese-educated biochemist Ephraim Racker first began coming out to talk, read, and make drawings and paintings of his fellow scientists. Also in 1946 the Berlin-born physical chemist Leonard Michaelis began living in Williams House, which he found to be less than luxurious. He became rather indignant about a crack in his apartment's toilet seat which Director Demerec apparently was reluctant to replace prematurely. Always conscious of the need to save money Demerec was frequently observed on the prowl turning off lights in seemingly unoccupied rooms, which was a source of much amusement to the summer visitors.

In addition to the sport made of Demerec's thrifty habits, other diversions from the long hours in the laboratories were walks to the Sand Spit for swimming and girl-watching, softball games played near Barbara McClintock's cornfield next to the Main Building, and canoe trips across the harbor to the village in pursuit of either ice cream sundaes or clams on the half shell. There was a tennis court to the north of Jones Laboratory, although its surface was rough and undependable. No one knew how to play golf except Richard Roberts from Carnegie's Department of Terrestrial Magnetism in Washington, who as part of a seminar series organized by the Department of Genetics gave a lecture in the summer of 1948 on "Extrasensory Perception" that was boycotted by Salvador Luria. Another talk in the series was presented by William Sheldon of the College of Physicians & Surgeons of Columbia University. Sheldon spoke on "The Natural History of Homo Americanus" in which he divided Americans into "happy/fat," "bland/normal," or "tense/thin" somatotypes. It is hard to understand why Demerec had invited him out to Cold Spring Harbor that summer since his chief occupation, apart from playing tennis, was the completion of a book entitled *Varieties of Delinquent Youths.*

Year-round research resumes at the Bio Lab

When the war ended the facilities in Nichols Building and James Laboratory were once again available for year-round research and this gave Demerec the opportunity to appoint more geneticists to the staff. Working in Nichols Building, Albert Kelner and Vernon Bryson used a grant from Schenley, the whiskey distillers, to continue the Bio Lab's successful wartime efforts to improve antibiotic yield by X-ray- and ultraviolet-light-induced mutations. Later, with support from the National Tuberculosis Society, Bryson focused on mutations that made the tubercule bacteria resistant to antibiotic therapy. In the course of his antibiotic work Kelner made the major discovery that visible light can partially reverse ultraviolet light damage to genes. This happened by chance when for seemingly inexplicable reasons he could not obtain reproducible results. As it turned out his results were being affected by the variability of the light in his ground floor laboratory in Nichols Building, which fluctuated greatly depending on whether the newly available fluorescent lights were turned on or off. Delbrück later remarked that Kelner's triumph reflected the "principle of limited sloppiness" and Kelner took offense at this slight to his talents.

James Laboratory became the site of studies funded by the Atomic Energy Commission on the effects of X-ray- and gamma-ray-induced mutations on the viability of populations of *Drosophila*. These experiments, which were led by Theodosius Dobzhansky's student Bruce Wallace, showed that some induced mutations increased viability, but it eventually became apparent that such population studies are virtually impossible to interpret. Experiments were also carried out by James King, who studied the emergence of *Drosophila* mutants resistant to DDT after long-term exposure to this insecticide.

(13) Waring blender used by Alfred Hershey.

His "blender experiment" proved that DNA is the genetic material. (Photographed at the DNA Learning Center of Cold Spring Harbor Laboratory.)

Carleton MacDowell retires and Alfred Hershey takes his place

In 1950 Carleton MacDowell was approaching 65, the mandatory retirement age for Carnegie Institution scientists, and to fill his slot Demerec brought Alfred Hershey from Washington University. Hershey was an early and influential member of the "Phage Group," the fraternity of bacteriophage workers led by Max Delbrück, who by then had been appointed to a professorship at the California Institute of Technology. Less than two years after his arrival to take over a second floor laboratory in the Animal House, Hershey, with his assistant Martha Chase, performed in 1952 the famous "blender experiment" in which he used the then innovative kitchen appliance in an elegant proof that DNA was the genetic material of the bacteriophage T2. Hershey later shared a 1969 Nobel prize with Luria and Delbrück for this landmark discovery in viral genetics. *(13)*

The Phage Meeting begins and a Bacterial Genetics course is added to the summer program

In August of the same year Hershey arrived, 1950, a Phage Meeting was held at Cold Spring Harbor for the first time. Thirty-five phage workers were in attendance, and at the end of the second day's talks Hershey summarized the science of the meeting, a tradition he would follow for many years. Carleton MacDowell called a square dance the first evening and a beer, pizza, and water-throwing party occurred the second evening, complete with a skit that featured a trial in which Delbrück was found guilty of spitting into his bacterial culture and sentenced to twenty years of hard labor at the California Penitentiary of Technology. Present at this meeting was the over-six-foot-tall Milanese aristocrat Niccolo Visconti de Modrone. Visconti did some experiments in Demerec's laboratory but later he worked with Delbrück to produce their mathematically elegant VD (Visconti-Delbrück) theory of how bacteriophages genetically recombine within bacteria. Visconti had also been one of 12 students in the new Bacterial Genetics course that Demerec had taught earlier that summer with Evelyn Witkin of the Carnegie staff and Vernon Bryson from the Bio Lab.

A new laboratory is erected for the Carnegie geneticists

Having moved into the post–World War II arena of molecular biology the year-round scientists of the Carnegie Institution's Department of Genetics were finding that the 1905 Main Building and the 1914 Animal House were inadequate for the kinds of research they now wished to do. As early as 1946 it had become apparent that a major new building would be required if the genetics program at Cold Spring Harbor were to have a future in a scientific world in which molecules as well as chromosomes were investigated.

To help finance the proposed new 16,000-square-foot laboratory building (later named Demerec Laboratory) Demerec eventually decided to sell the Eugenics Record Office (ERO) and the surrounding acreage bordering on Stewart Lane. For a brief time after becoming director he had considered inaugurating a serious human genetics effort to replace the racially tinged, and finally defunct, eugenics program of the Carnegie Department of Genetics. In the summer of 1941 he had offered a position to James Neel, a recent Dartmouth Ph.D. working with *Drosophila* who wanted to move into human genetics. When Neel decided that his future career might require a medical degree and headed off to medical school instead of coming to Cold Spring Harbor, Demerec could find no other suitable candidate for his proposed new human genetics program. Left with no real use for the ERO facility and in need of funds for the construction of a modern laboratory structure on Bungtown Road, Demerec finally put the ERO land and buildings up for sale. (The records were removed to the Dight Institute of Human Genetics of the University of Minnesota at Minneapolis, where they remain to this day in the Institute's basement.) Part of the ERO land was sold to West Side School and the remainder became house plots, several of which were purchased by Amyas Ames, who later became president of the Long Island Biological Association (LIBA). Before construction could begin on the new laboratory, however, a perennial form of Bio Lab housing needed to be moved.

CABINS
(Built in 1930s and 1940s)
Relocated
1951

(14) Tents on Blackford Hall lawn and "Tenters' Anthem."

Before there were wooden cabins for summer visitors there were wood-framed canvas tents. Pity the students who did not have rooms in Blackford Hall or one of the old homes used as dormitories. Their trials are enumerated in the "anthem" (sung to the tune "Funiculi, Funicula") accompanying the photograph.

For many years wood-framed canvas tents had been a regular feature of the summer landscape at the Bio Lab and these had been followed by simple wooden cabins. In the fall of 1951 the time came for the cabin colony on the lawn behind Blackford Hall to be disbanded to make way for the Carnegie Institution's large construction project that was about to commence directly next door. In the time-honored Bio Lab tradition of "waste not, want not" the cabins

were not torn down but instead moved to a site on the other side of Bungtown Road. In their new location on the hill behind James Laboratory, not far from Urey Cottage, they were recombined into four unique "guest cottages" and given names—Cabin A, Cabin B, Cabin C, and Cabin D—recalling their rustic past. These upgraded cabins would for many years continue to provide much needed housing for summer visitors. *(14,15,16)*

Some think this world was made for cozy luxury,
But not so I; but not so I!
If but a gently leaking tent they'll give me
I'll never cry; I'll never cry.
I love to sink upon the yielding corncobs
That form my bed, that form my bed,
And feel the rain upon my cran-i-al knobs,
From overhead, from overhead.
    Tenting, tenting in the freezing dawn
    Brush your teeth with the neighbors looking on,
    The while they hear you stamp and cussing at
    The millions of the bites—
    Where the scientifical bugs
    Collected you of nights.
Some think that oily lamps died with the old ark,
But they don't know, but they don't know
How nightly I seek matches when the tent's dark,
And stub my toe, and stub my toe;
Nor how that lamp doth flare and die each second,
Eccentric'ly, eccentric'ly
With all the beetles, bugs and skeeters beckoning
To sleep with me, to sleep with me.
    Midnight tenting bi-o-log-i-c'ly;
    While a storm howls, with hand lens on my knee,
    I sit with sci-en-tif-ic in-ter-est
    A-waiting just to see—
    When the whole tent blows away
    What will be left of me.
(From "Cold Spring Harbor Songs," undated)

(15) Cabins on the lawn between Blackford Hall and the Animal House (now McClintock Laboratory) circa 1935.

These cabins superseded the canvas tents.

(16) Cabin D today.

When construction began on Bush Lecture Hall and Demerec Laboratory in the early 1950s the wooden cabins had to be relocated to a site on the hill on the west side of Bungtown Road near Urey Cottage. Here they were combined in twos and threes to form four "guest cottages" (the name did not stick). In the late 1980s when construction of the 1991 Neuroscience Center was under way two of these cabins were removed, but Cabins C and D still remain adjacent to the north entrance to the Center.

(17) Foundations of Demerec Laboratory being poured, October 1951.

The former Bio Lab director's residence (now Davenport House) can be seen in the background.

DEMEREC LABORATORY
Anderson & Beckwith
1953

Once ground was finally broken, construction proceeded quickly on the Carnegie Institution's multistoried concrete laboratory complex consisting of two building sections nestled into the hillside that slopes down to the harbor on the east side of Bungtown Road. The structure was designed by the Boston architectural firm of Anderson & Beckwith in the form of two parallel, interconnected, oblong masonry boxes punctured evenly with large plate glass windows. One section was positioned closer to Bungtown Road than the other, but both turned their short sides in that direction. The main entrance to the complex, located under a cantilevered canopy of concrete, was situated at the edge of a passenger drop-off loop. This Bungtown Road entrance was on the top floor of the more southern three-story section of the building that was closest to Bungtown Road and New York State Route 25A. Because the land falls off steeply on the east side of Bungtown Road near its intersection with Route 25A, only the top floor on this main section of the building was on a level with the road. *(17, 18, 19)*

(18) Walls of Demerec Laboratory going up, November 1951.

The 1905 Main Building (soon to become Carnegie Library) is visible on the right in the background.

(19) Front view of Demerec Laboratory nearing completion, October 1952.

(20) Rear view of Demerec Laboratory as completed, May 1953.

The other section of the structure was situated to the north and set farther back from the road. Only two stories tall, this smaller section was connected to the main section by a stair tower located between the two at the back of the building which faced the water. The tower also provided immediate access to both sections through a ground floor door situated on its harbor side. (20,21)

Completion of this new structure, which was later named Demerec Laboratory, freed up the 1905 Main Building to become a true library facility. In addition to housing laboratories, the Main Building had long served as the main repository of reference works and journals for both the Carnegie Department of Genetics and the Bio Lab, and with this now its sole function it soon became known as Carnegie Library.

BUSH LECTURE HALL
Anderson & Beckwith
1953

Designed in light, airy Scandinavian Modern style, the new bi-level lecture hall was built into the hill between the new Carnegie laboratory then under construction and Blackford Hall. It not only contained a 3300-square-foot Lecture Room with a capacity for 250 folding-chair seats, but also incorporated both a wood-paneled lounge (now called the Fireplace Room) and a breezeway under its long low roof that sloped dramatically downward in the direction of the harbor on its east-facing (rear) facade. *(23)*

Like the neighboring laboratory the lecture hall was constructed of reinforced concrete, but it was then sheathed in

(23) Lecture Room of Bush Lecture Hall, 1953.

This photograph shows all of the important features of the acoustically advanced design of the Lecture Room—"no parallel surfaces, no vertical walls, wave-shaped ceiling, and baffled rear wall" (quoted from *Carnegie Institution of Washington Yearbook* for 1953).

warm reddish-brown brick and topped with a cupola. At the dedication ceremony held for both new facilities at the start of the 1953 Symposium, newly installed LIBA president Amyas Ames, who later served as chairman of Lincoln Center, drew the audience's attention to the acoustically advanced elements of the Lecture Room's design—"no parallel surfaces, no vertical walls, wave-shaped ceiling, and baffled rear wall." The building was later named in honor of Vannevar Bush, the Carnegie Institution of Washington president who was the single greatest proponent of the expansion of scientific research in the United States in the years following World War II. *(24,25)*

(24) Bush Lecture Hall in the mid-1960s.

The 250-seat Bush Lecture Hall served as the scene of the Cold Spring Harbor Symposia on Quantitative Biology from 1953 (when the meeting was moved here from the Assembly Room of Blackford Hall, capacity less than 100) until 1986 (when it moved into the new 360-seat Grace Auditorium).

(25) Bush Lecture Hall today.

Its Fireplace Room continues to be used for small presentations and the former Lecture Room is used for poster sessions during meetings.

Finding Out How Genes Work

Genes control the properties of cells by specifying which proteins they possess. In 1909, soon after the rediscovery of Mendel's laws, the English physician Archibald E. Garrod linked several hereditary diseases to the absence of specific enzymes involved in the breakdown of specific amino acids found in our food. In time additional biochemical defects caused by gene mutations were found, culminating in George Beadle and Edward Tatum's use in 1940 of the mold *Neurospora* to show that the successive steps in the biosynthesis of amino acids and vitamins were each controlled by gene-specified enzymes. The protein nature of all then-known enzymes had been established, with each enzyme, like all other classes of proteins, consisting of a linear collection of amino acids known as polypeptide chains. Twenty different amino acids exist, with each specific polypeptide chain having its own specific amino acid sequence whose length varies from one protein to another, ranging from as few as fifty amino acids to more than several thousand. Many proteins contain only one polypeptide chain, in which case they are specified by single genes. In other examples several genes must cooperate to provide the information for a protein. The blood protein hemoglobin is made up of two polypeptide chains, each encoded by a different gene.

As soon as the double helical structure of DNA was found the gene-protein relationship was rephrased as a coding problem. How do the sequences of bases along DNA molecules determine (code for) the order of the amino acids in their polypeptide products? From the start it was clear that DNA sequences do not directly determine amino acid sequences. Instead the DNA base sequence information is copied (transcribed) into the sequences of the second form of nucleic acid, RNA (ribonucleic acid). The RNA base sequences in turn serve as molds (templates) on which the amino acids are assembled into polypeptides. Chemically RNA is very similar to DNA, also containing four bases, three of which (A, G, and C) are identical to those of DNA. Its fourth base, U, is closely related to the T of DNA and forms A-U base pairs. RNA chains are usually single-stranded and as such serve as templates for protein synthesis.

Because many more amino acids exist than RNA bases, a number of bases must specify (code for) a single amino acid. That this number is three was first shown in 1961 by Francis Crick and Sydney Brenner at Cambridge. They used mutations in a bacteriophage gene to show that successive groups of three bases (a codon) along RNA templates specify successive amino acids in their polypeptide products. The first codon identified was UUU, which codes for the amino acid phenylalanine. Most amino acids were found to be encoded by more than one codon, with 61 of the potential 64 (4x4x4) codons used to specify amino acids. The remaining three codons convey stop signals that terminate polypeptide assembly. The list of codons and the amino acids they specify is known as the "genetic code." Established by 1966, only a decade after the search for it began, the genetic code is effectively the same for all forms of life. For example, once we know the DNA sequence in any bacterial gene the amino acid sequence of its polypeptide product is also known. The flow of genetic information is thus:

$$\text{DNA} \xrightarrow{\text{(transcription)}} \text{RNA} \xrightarrow{\text{(translation)}} \text{protein}$$
(duplication)

Why such a seemingly tortuous pathway exists to make proteins was obscure for almost three decades. Then in 1981 Thomas Cech found that some RNA molecules are enzymes and, similar to many proteins, can greatly speed up (catalyze) the rates of specific chemical reactions. This major discovery led to the suggestion that when life started on earth the first genetic molecules were composed of RNA, not DNA. Then proteins did not need to be present since the linking together of bases to make new RNA molecules could be catalyzed by RNA itself. Protein synthesis need only have evolved after the *RNA world* had begun. After the start of protein synthesis a specific protein enzyme that makes DNA using RNA templates must have evolved. In time DNA became the primary genetic molecule because DNA chains are chemically more stable than RNA chains and so are capable of existing as much longer molecules. Today RNA exists in nature as truly self-replicating molecules only as the "chromosomes" (genetic molecules) of certain viruses. These RNA viruses may in fact be molecular relics of the much earlier RNA world.

*Cold Spring Harbor is THE summer destination for
molecular biologists*

The 1953 Symposium on Viruses, the first to be held in the new Bush Lecture
Hall, was a major event in the history of molecular biology and genetics. It was the first
large public forum at which the startlingly simple idea that the DNA molecule is a gently
twisted ladder was discussed. The news was brought straight from Cambridge University by
geneticist James Watson, who together with Francis Crick had just elucidated the double
helical structure of the molecule of heredity, DNA (deoxyribonucleic acid). For this
discovery Watson and Crick later shared a 1963 Nobel prize with Maurice Wilkins. This was
not Watson's first trip to Cold Spring Harbor. He had first come to the Bio Lab in the
summer of 1948 as a graduate student of Salvador Luria to do experiments in Nichols
Building and Davenport Laboratory, and in August 1950 he had stopped at Cold Spring
Harbor on his way to a postdoctoral fellowship in Europe to participate in the Phage
Meeting. Later, in 1968, he would return as the new director of the Cold Spring Harbor
Laboratory of Quantitative Biology.

While the Symposium was continuing to grow in size the number of courses taught
at the Bio Lab during the summer was also increasing. A Genetics of Fungi course was offered in 1956 and 1957 and Herman Moser of the Bio Lab started an influential course on
Animal Cell Culture (later broadened to include animal viruses) in 1958. Among the first 14
students in the Animal Cell Culture course was Purnell Choppin, who after a distinguished
career at Rockefeller University later became president of the Howard Hughes Medical Institute.

With this growing influx of scientists and students to the Bio Lab each summer
more on-site housing was needed. The vast majority of summer visitors arrived by train or
plane, and local mass transportation was, and still is, virtually nonexistent. Fortunately the
campus itself had expanded through an odd sequence of events that occurred during the period that Demerec Laboratory and Bush Lecture Hall were nearing completion. It was a
tribute to the perseverance of Director Demerec that the Bio Lab had finally come out ahead
in protracted negotiations with neighboring landowner Rosalie Jones, a very formidable op-
ponent.

*Rosalie Jones sells property on Bungtown Road to the
Bio Lab*

Rosalie Gardiner Jones (1883–1978) was the daughter of Oliver Livingston Jones.
As a young man Oliver Jones had bought and developed one of Cold Spring Harbor's
premier resort complexes, the Laurelton Hotel (see Chapter One), and over the years he had

acquired extensive interests in local real estate. Rosalie Jones grew up in a Jones family mansion, long known as the Manor House, on the east side of Syosset-Cold Spring Harbor Road. Built in 1855 by Walter R. Jones, the financier of the "Jones Industries" and founder of the Atlantic Mutual Insurance Company, the house was inherited by Miss Jones' grandfather Charles H. Jones (youngest brother of Walter R.) and later by her mother. After the Manor House was destroyed by fire in 1909, Rosalie's father had an equally imposing stuccoed residence built which she later inherited (it is now a nursing home).

Rosalie Jones was at one time married to Clarence Dill, a United States senator from Spokane, Washington, and was herself a lawyer, having graduated from Adelphi College and Brooklyn Law School. Shod in her signature white tennis shoes, she was a familiar sight on the platform at the Cold Spring Harbor railroad station enroute to the Madison Avenue legal firm where she worked in New York City. Her obituary notice in the Long Island daily newspaper *Newsday* on January 13, 1978, headed "Rosalie Jones, 95, An Early LI Feminist," noted that she was born "at a time when women in wealthy families were expected to stay home and paint china, but she chose instead to march with the suffragettes.... The experience put her out of step with most other debutantes of her day and she remained a bit out of step with her native Cold Spring Harbor." This was never more evident than when she rented an old home she owned on the east side of Bungtown Road just north of then Davenport Laboratory (now part of Delbrück and Page Laboratories) to a succession of migrant workers and their families. Long known as "Rosy's Cozy," this house burned to the ground in the mid-1960s.

When in the early 1950s Rosalie Jones threatened to develop as half-acre plots her twenty-two-and-a-half acres of prime waterfront property next door to the Bio Lab, Director Demerec went into action. According to the minutes of an early 1953 meeting of the LIBA board, Demerec was determined to "insure the privacy that is essential for efficient operation of the laboratories." After months of negotiations, agreement was finally reached to purchase the property. It was also agreed that to raise the necessary funds LIBA would sell eight acres of the former Townsend Jones land bordering on Moore's Hill Road for development as four house plots. At the same time LIBA adopted a resolution consenting to the annexation of all of the Bio Lab property to the Incorporated Village of Laurel Hollow where by law house plots had to be at least two acres in size. The four new lots that LIBA sold were immediately adjacent to the half-acre site where Hugo Fricke had built his small Henry Saylor-designed cottage in the late 1920s. Among the purchasers of these lots was young lawyer James A. Eisenman, who in addition to his duties as longtime village justice of Laurel Hollow was to serve for nearly twenty-five years as treasurer of LIBA.

In 1955, two years after the negotiations with Rosalie Jones and the Village of Laurel Hollow were concluded and the building lots had been sold, the Bio Lab decided to make use of an additional part of the Townsend Jones land, a level section high on the wooded west side of Bungtown Road, for the construction of a sorely needed "motel" facility.

PAGE MOTEL

George B. Post & Sons
1955 and 1959
(Demolished in 1987)

Situated just south of Urey Cottage and the Cabins, the first wing of the motel was erected in 1955. A second wing went up in 1959. Consisting of 18 unheated double rooms, the two wings formed an L-shaped structure in typical single-story motel style, complete with knotty pine interiors. The complex was named Page Motel in honor of Arthur W. Page, who had retired from the LIBA board in 1951 after more than twenty-five years of service both as treasurer and president.

Life at the top of the Bio Lab at Page Motel was most convivial, with bar-becues and lawn chairs very much in evidence on the large grassy courtyard created by the two wings. Sadly Page Motel had to be demolished in 1987 after the broad level site that it occupied halfway up the hillside was deemed the best available place for a Neuroscience Center to be constructed (completed in 1991). In the same year that the Motel was demolished, however, a brand-new facility bearing the name Page was dedicated—the Arthur and Walter Page Laboratory for research and teaching in plant genetics (see Chapter Six). *(26)*

(26) Page Motel in the late 1960s.

Erected in two sections in 1955 and 1959, Page Motel consisted of 18 double rooms. The section completed in 1959 was graced by a cupola (partially obscured in this photograph). The Motel was demolished in 1987 to make way for the Neuroscience Center (completed in 1991) which incorporates a new 60-room visitor housing facility, Dolan Hall.

(27) Hooper House today.

The former ground floor entrance and flanking sidelights were transformed into one broad window during renovations in 1960.

New big government grants help support the summer program and renovate facilities

The year after the second wing of Page Motel was completed its architects George B. Post & Sons of Huntington were given another commission related to housing at the Bio Lab. After almost the entire interior of circa 1835 Hooper House was gutted in 1960 a new floor plan from Post & Sons was executed. Several comfortable year-round apartments were created on the ground and first floors, and the top floor was reconfigured as a seven-room women's dormitory. The firm also furnished plans for adding a second floor to James Laboratory for bacterial virus work and these plans were executed in 1961. All of these improvements as well as renovations to Blackford Hall to improve cafeteria service were made possible by a $135,000 four-year grant from the National Science Foundation. Also invaluable at this time was a five-year National Institutes of Health grant of $14,000 per year to support the summer training courses at the Bio Lab. *(27,28)*

(28) James Laboratory today.

Like Page Motel and the renovations to Hooper House, the second floor addition to James Laboratory built in 1961 was designed by George B. Post & Sons of Huntington. Subsequent additions to James Laboratory would carry through the vertical board exterior sheathing motif introduced here. James Annex (built in 1971) can be seen abutting it on the south (*left foreground*). James Laboratory functions today as a tumor virology laboratory.

*Siamese fighting fish provide a nonhuman assay
for LSD*

In the summer of 1961 psychiatrist Harold Abramson performed the last experiments at the Bio Lab in his ongoing study of psychotropic drugs. He had previously been at Cold Spring Harbor in the late 1930s to do research on proteins in Nichols Building, but after the war he had become interested in the newly discovered psychotropic drug LSD, whose administration to humans provoked schizophrenia-like delusions. He participated in experiments at Mt. Sinai Hospital in which LSD was given to human volunteers, and for many years at Cold Spring Harbor he studied its effect on Siamese fighting fish and carp. Abramson, who lived nearby on Stewart Lane in Laurel Hollow, was a somewhat controversial member of the local community and no one was surprised when in 1982 in an unexpected bit of candor the Central Intelligence Agency revealed that it had provided the funds for the LSD research, which by then had largely been forgotten.

Demerec prepares for his retirement from Carnegie

About to turn 65 in 1960, Director Demerec knew that no exception would be made to the obligatory Carnegie Institution retirement age but he hoped to continue his research on bacterial genetics by moving over to the Bio Lab. In 1958 he persuaded LIBA to hire an architect to plan a new laboratory since both Nichols Building and James Laboratory were by then filled to capacity. The work being done in these laboratories was constantly improving, and even though Bruce Wallace was moving on to become a professor at Cornell University, another talented *Drosophila* scientist, Arthur Chovnick, was coming from the University of Connecticut. The very productive bacterial geneticist Ellis Engelsberg had left for the University of Pittsburgh, but he was replaced by Indiana University-trained Paul Margolin. There was still the problem of having no endowment with which to support salaries, however, and even though LIBA announced a fund-raising goal of two million dollars, it was an unrealistic objective, particularly since the community assumed that the Bio Lab's stability was ensured by the large endowment behind the Carnegie Department of Genetics. When in fact it proved impossible to raise the monies for the proposed new laboratory, plans were made instead for the second floor addition to James Laboratory (see illustration *28*) that was completed in 1961 using funds from a government grant.

With his retirement plans still uncertain, Demerec was faced with yet another problem, albeit one of a less personal nature. The Town of Oyster Bay proposed to dredge its portion of the inner harbor, the western side where the Bio Lab was situated (the county line runs down the middle of the harbor). If that happened, boats would be able to moor right next to the Laboratory instead of being forced to anchor some thousand feet away to avoid

(29) View of the inner harbor with swans.

In the late 1950s Director Milislav Demerec successfully headed off an attempt by the Town of Oyster Bay (backed by boating enthusiasts) to dredge the western side of Cold Spring Harbor.

being grounded at low tide. Earlier, Demerec had foiled an attempt to make the Sand Spit a town public beach and once again he prepared to do battle. He even sent letters to a number of prominent geneticists asking them to write to the supervisor of Oyster Bay. Among those who responded was J.B.S. Haldane, the most brilliant of all English geneticists and an old Etonian as well as a vocal Communist. His letter, which began "Dear Comrade," might well have been all that was needed to convince public officials and those in the yachting world who backed the dredging proposal that it would be better to keep some distance between themselves and the scientists. *(29)*

Demerec's triumph in preventing the dredging was soon counterbalanced by defeat when it became clear that there would be no laboratory space available for him at the Bio Lab after his retirement from his Carnegie position. By the time Arthur Chovnick assumed the directorship of the Bio Lab in 1960 he had already recruited Edwin Umbarger from Harvard Medical School to occupy the soon to be completed second floor of James Laboratory. Fortunately upon his retirement Demerec was able to move his research to the Brookhaven National Laboratories where he was given resources commensurate with his still very real talents. He continued to live in the Cold Spring Harbor area, having built a home off Stewart Lane on one of the plots of land that had formerly belonged to the Eugenics Record Office.

The Carnegie Institution of Washington casts off

In his search to find someone to replace Demerec as director of the Department of Genetics at Cold Spring Harbor, Caryl Haskins, then president of the Carnegie Institution of Washington, offered the position successively to three prominent molecular biologists— Italian-born Renato Dulbecco, who worked on animal viruses at the California Institute of Technology; South African-raised Sydney Brenner, who was working on the genetic code with Francis Crick; and bacterial geneticist Norton Zinder from Rockefeller University. All in turn declined the position. There were two reasons for these rejections. First, the position being offered was a less powerful one than Demerec had held because the Bio Lab now had its own director (and he, not the Carnegie director, would be living in the director's house, Airslie). Second, the Carnegie Institution did not want its departments to be dependent on federal funding and in fact discouraged its staff from seeking such funds. Vannevar Bush, Haskins' predecessor, had articulated the Carnegie Institution position in his speech at the Demerec Laboratory dedication ceremony when he expressed his fears that government subsidies would result in federal control over individual scientific efforts. He realized that to avoid this potential danger the Carnegie laboratories might have to forego elaborate or expensive equipment but noted that "the day is not yet passed when men with brains and very simple equipment and simple methods can accomplish great things." As a result of this Carnegie Institution philosophy of independence from outside help, which Haskins enthusiastically espoused, it would be impossible for the new director of the Department of Genetics to face head-on the intellectual challenges posed by the discovery of the double helix—exactly the kind of research for which federal science monies were then readily available.

After three rejections, Haskins in desperation proposed that Ellis Bolton, a Carnegie scientist who studied protein synthesis in bacteria at the Carnegie Biophysics Laboratory in Washington, become the new director. American geneticists soon protested, arguing that since Bolton was not a geneticist he was therefore unqualified for the position. Angered by this rebuff, Haskins made the decision to close down the Department of Genetics and hand over its physical resources to the Bio Lab provided it could become a financially viable organization. If this could be accomplished the Carnegie Institution would continue to support

Barbara McClintock and Alfred Hershey at Cold Spring Harbor. While negotiations with the Bio Lab were proceeding, Berwind Kaufman, who for many years had collaborated with Demerec in studies of chromosome structure, was made acting director of the Carnegie Department of Genetics in 1960. During his two-year tenure Kaufman greatly increased security at the library by having the head of the library, Guinevere Smith, lock its doors upon her departure each day at 5 p.m. This was an unprecedented occurrence at Cold Spring Harbor and totally out of synchrony with the scientists' lives. *(30)*

(30) Demerec Laboratory and Bush Lecture Hall at dusk.

Science at Cold Spring Harbor happens around-the-clock.

Walter Page helps to create the Cold Spring Harbor
Laboratory of Quantitative Biology

The effort to merge the facilities that the Carnegie Institution was abandoning with those of the Bio Lab was led by the then president of LIBA banker Walter Page, the son of an earlier LIBA president, Arthur Page, and a neighbor of the Bio Lab. (While ascending the corporate ladder to the top of Morgan Guarantee Trust Company the younger Page had built a home in Cold Spring Harbor that looked across the inner harbor to the Bio Lab.) Initially there was talk that the Associated Universities, the consortium of major east coast universities that operated the physics-oriented Brookhaven National Laboratories might also provide umbrella support for the geneticists at Cold Spring Harbor. This proposal was met with resistance from some biologists who insisted that biological research should not be under the control of physicists and that the new world of molecular biology should have its own group of university sponsors. It finally took two years of negotiations between the Carnegie Institution and LIBA, aided by its Scientific Advisory Board led by Edward Tatum of the Rockefeller University, to establish the Cold Spring Harbor Laboratory of Quantitative Biology (CSHLQB) which was formally incorporated in 1963. (The name was later simplified to Cold Spring Harbor Laboratory, CSHL, in 1970.) To guarantee the fiscal stability of the CSHLQB a select group of Participating Institutions, both universities and research institutes, made pledges to donate $25,000 each if the new organization were ever to be in financial difficulty. Thus control of the newly formed CSHLQB effectively lay in the hands of the Participating Institutions, which provided nine of the original dozen trustees.

The first director appointed by the CSHLQB trustees was Oxford-born M.D. and molecular biologist John Cairns. Although he had been working in Australia, Cairns was not a stranger to Cold Spring Harbor since he had earlier spent a sabbatical year as a visiting scientist in Alfred Hershey's laboratory. During his stay he had elegantly shown that the chromosome of bacteriophage T4 was a single DNA molecule. Upon his return to Canberra Cairns went on to prove that the much larger chromosome of *E. coli* was also a single DNA molecule in which the two ends had fused to form a circle. The Cold Spring Harbor Laboratory of Quantitative Biology was most fortunate in having acquired as its first director an intellectually worthy successor to Demerec.

John Cairns rescues the physically decrepit, financially
bankrupt CSHLQB

Upon his arrival at Cold Spring Harbor in July 1963 John Cairns took over a physically run-down research institute with more debts than free cash. During the brief interval between Demerec and himself there had been a great outpouring of resources with little record of where they went. Cairns immediately obtained a grant from the National Science Foundation to assess how much it would cost to renovate the CSHLQB laboratories and residences. The answer was $750,000, a sum that greatly exceeded the Laboratory's immediate potential to raise money from the local community or its scientific friends. Thanks to the sales of the increasingly popular *Symposia* volumes, however, the Laboratory's actual cash position gradually moved from a negative to a positive one. Some of the surplus cash went into badly needed maintenance, including new roofs for both the 1914 Animal House (now McClintock Laboratory) and 1907 Blackford Hall, which had its walls and balcony repaired as well. Also very helpful were sizable donations from a number of pharmaceutical companies as well as the Selman Wakesman Foundation for Microbiology. Equally important was the Rockefeller Foundation commitment to cover Cairns' salary for five years.

Cairns pursued his own research installed in a former Carnegie facility, the still new Demerec Laboratory where Hershey and McClintock now worked. He quickly strengthened the scientific staff by bringing Cedric Davern from Australia and Joseph Speyer from New York University. Both of these scientists soon had sizable grants, as did Cairns, and a small but real cash surplus accumulated. More important scientifically, their work helped shift the Laboratory's main intellectual thrust away from genetic crosses and toward the use of molecular approaches to understand how DNA replicates and functions.

Cold Spring Harbor Symposia volumes become the
"bible" of molecular biology

At first there were no texts to document the new world of molecular biology, but beginning with the 1946 meeting on Heredity and Variation in Microorganisms this void began to be filled by *Symposia* volumes. Papers presented at the 1961, 1963, and 1966 Symposia recorded the search for the genetic code; in fact, the 1966 volume chronicled the announcement of its solution which was presented for the first time at that year's Symposium on The Genetic Code (at which, coincidentally, Francis Crick's 50th birthday was also celebrated). As early as 1961 the interest in molecular biology was so great that the Symposium audience far exceeded the capacity of Bush Lecture Hall and closed-circuit television was installed in the adjacent Fireplace Room as well as in the bar in the basement of Blackford Hall. Tents erected on the lawn behind Blackford Hall provided extra dining facilities, and mealtimes were noteworthy for the long lines in the cafeteria. *(31,32)*

Even after the genetic code was solved there was still no time to relax intellectually at Cold Spring Harbor. The very next year the world's leading immunologists congregated at the 1967 Symposium on Antibodies to discuss the apparent paradox that many more antibodies exist than there are different genes to code for them. Macfarlane Burnet arrived from Australia to open the meeting with a talk on "The Impact of Ideas on Immunology," and

(31) Blackford Hall lawn, scene of Symposium meal service.

The scene captured in this photograph has changed. By the Laboratory's centennial year 1990 the number and sizes of the yearly meetings at Cold Spring Harbor had increased to such an extent that an addition to the back of Blackford Hall was begun.

(32) Sketch of Blackford Hall made in 1951 by Alexander Kohn.

A tongue-in-cheek allusion to the meals served to summer visitors, this drawing by a participant in the Phage Course is labeled "Depletion Experiments," a play on words since this also describes a type of experiment performed in the summer courses.

DEPLETION EXPERIMENTS

(33) Airslie lawn, scene of Symposium picnics and soft–ball games.

Each year the Symposium participants are treated to a bountiful picnic down by the beach.

Nils Jerne from the Paul Ehrlich Institute in Frankfurt closed the meeting with a summary entitled "Waiting for the End." When possible, relaxation at the Symposia in the 1960s meant the time-honored picnics and softball games on the lawn of Airslie, the latter all too often interrupted by balls disappearing into the hard to eradicate poison ivy patches that had sprung up in the 1950s after the dissolution of Henry de Forest's estate. Wonderful respites each year were the Sunday evenings when the Symposium speakers went to the homes of LIBA members for dinner parties that brought them together with prominent figures in the local community. *(33)*

James Watson, a professor at Harvard, simultaneously takes on the directorship of the CSHLQB

Despite the intellectual triumphs of his Symposia, Cairns worried that his institution could not survive if it failed to acquire an endowment to cover the salaries of its key scientists. Once Hershey and McClintock were gone there would be no further Carnegie monies to support the salaries of distinguished replacements. To his dismay both Speyer and Davern, the two scientists he had recruited, decided to move on to potentially tenured university positions. Cairns saw no alternative but to resign unless he got more help from his Board of Trustees.

Relations between Cairns and Edward Tatum, the chairman of the CSHLQB board, had been tense from almost the very beginning. Tatum convened all of the board meetings at his home institution, Rockefeller University, and did not make a single personal visit to the Laboratory after Cairns became director. When the CSHLQB was being created Tatum had relied heavily on the support given him by Francis Ryan, the first vice president of the LIBA board. It was Ryan, not Tatum, who solicited letters from a host of Nobel laureates in support of the new CSHLQB and who had first suggested to Cairns that he become the new director. When just six weeks after his arrival at Cold Spring Harbor Cairns wanted to arrange for at least one board member to make a proper tour of the site so that the true predicament of the Laboratory would be known to the trustees, it was to Ryan that he turned for help. Ryan came out from New York the next day and was so horrified by what he saw that he promised to telephone Tatum to suggest that he immediately convene a meeting of the board. Tragically this energetic CSHLQB advocate died of a heart attack the next day at the age of 47. No meeting of the board took place that summer, and for the remainder of Cairns' tenure as director of the CSHLQB, the Board of Trustees, continuing to meet in Manhattan, made no significant contribution to the Laboratory on Long Island. To make matters worse, the usually steadfast support given by the local community had diminished during this period. Many LIBA members were still upset that control of the Laboratory had been given up to absentee university trustees who had no emotional ties to the site or its long history. After four years of carefully managing the Laboratory's limited resources and trying to raise funds from a less than receptive audience Cairns decided to resign.

Upon receiving Cairns' resignation in early 1967 Edward Tatum did little to persuade him to change his mind. Instead, he and fellow trustee Irwin Gunsalus were keen to offer the position to Carsten Bresch, a German bacteriophage worker whose main qualification was his desire to leave his position in Dallas. However, at the fall trustees meeting an alternative was offered by James Watson, who had joined the board several years before at Cairns' request. Watson suggested that the dilemma of how to pay the director's salary would vanish if he were to become the director while remaining a full-time professor at Harvard. Bentley Glass, the Stony Brook geneticist who had just succeeded Tatum as chairman of the board, later followed up on the idea. After consulting with fellow board members, Glass formally offered the nonsalaried position to Watson, who in turn sought formal consent from Harvard. This was given without objection so long as Watson continued to tend to his teaching and departmental duties there. In February of 1968 Watson officially became the director of the Cold Spring Harbor Laboratory of Quantitative Biology and he immediately asked Cairns to stay on as a member of the staff and to continue living in Airslie. Watson's main residence was to remain in Cambridge, Massachusetts, on Appian Way near Harvard Square. For their Cold Spring Harbor residence Watson and his recent bride Elizabeth chose Osterhout Cottage, the tiny early nineteenth century house that Charles Davenport and his family had occupied when they first arrived at the Bio Lab in the late 1890s and where Alfred and Jill Hershey lived when they first came to Cold Spring Harbor in 1950. *(34)*

(34) Osterhout Cottage in the early 1950s.

Alfred Hershey and his family were living in this early nineteenth century house at the time.

Reconstruction of
OSTERHOUT COTTAGE
Edelman & Salzman
1969

To accommodate the new director and his wife, plus future family, plans were made to renovate and expand the circa 1800 story-and-a-half shingled cottage that was tucked into the bend in Bungtown Road just north of Blackford Hall. Right after the start of the project, however, the builder decreed that the original structure could not be saved and proposed building to the same architectural specifications an entirely new air-conditioned house for the same price previously agreed upon for the renovation.

The cottage was thus rebuilt according to what were originally renovation designs by partner Harold Edelman of Edelman & Salzman, the New York City architectural firm that had provided a space use study and master plan for the Laboratory in the mid-1960s. Now restructured (since 1979) as the Edelman Partnership, the archi-

(35) Osterhout Cottage today.

In 1968 new Laboratory director James Watson, still a professor at Harvard, chose Osterhout Cottage as his Cold Spring Harbor base. In 1969 what started off as a renovation project turned into a reconstruction one due to the fragile state of the house. It was completely rebuilt with a tall new wing and as a bonus the builder installed central air conditioning, nice to have on steamy July nights.

tects of the Osterhout renovation (and several later projects at the Laboratory) maintain a general practice (designing co-op and luxury apartments, loft renovations, and office interiors) as well as specialize in two areas: historic restorations and subsidized housing commissioned by not-for-profit sponsors such as community groups. In the former category they are especially noted as the designers of the 1975–1978 restoration of historic 1799 St. Mark's-in-the-Bowerie Church in the East Village section of New York City. Following a catastrophic fire in 1978 that gutted the newly restored church it was entirely rebuilt (1978–1984) by the same Youth Corp manpower that had been assembled for the initial restoration using the extensive original restoration drawings.

The earlier case of Osterhout Cottage at Cold Spring Harbor thus bears a certain similarity to that of St. Mark's-in-the-Bowerie in that working drawings originally prepared for a straightforward renovation project became the basis for entirely rebuilding a structure from scratch. As rebuilt, the lines of Osterhout Cottage on its south side facing Blackford Hall were identical to those of the original structure except for the floor-to-ceiling shed-roofed dormers with railings that were inserted on the second floor of the house both front (facing Bungtown Road) and back (overlooking the harbor). An earlier addition at the north end of the original cottage was rebuilt into a two-story wing with a vertical slit window. *(35)*

With its unusual fenestration (including bubble skylights on the east-facing roof slopes) the new cottage completed in the summer of 1969 bore only a remote resemblance to the original, but the potentialities of the site for panoramic views of the inner harbor with its abundant waterfowl had been utilized to the utmost. The funds for the project were donated by the Laboratory's new director, whose 1968 account of his prize-winning discovery, *The Double Helix,* had proved to be a best-seller.

The future of the Laboratory begins to look brighter

Long before the plans for renovating Osterhout Cottage were formulated, however, Director Watson's first concern in 1968 was to find funds that would allow John Cairns to remain at Cold Spring Harbor. Fortunately the following year the American Cancer Society awarded Cairns one of its prized professorships, effectively guaranteeing his salary for as long as he wished to stay at the Laboratory. Thus by the end of 1969 the maintenance of real science along Bungtown Road was assured even though the Laboratory would be sharing its new director with Harvard.

CHAPTER FIVE

Cold Spring Harbor Laboratory adaptively reuses its Bungtown heritage

1970~1980

When James Watson became the director of the Cold Spring Harbor Laboratory of Quantitative Biology in 1968 he recognized that one route toward achieving fiscal security for the Laboratory would be to create a major new research objective that had the optimum potential for government funding. The research problem he chose was a challenging one—how viruses cause cancer—and soon afterward the president of the United States himself declared a national "War on Cancer."

In the midst of an explosion of highly successful tumor virus research the old world of genetics was not forgotten. Molecular biologists seeking to understand how genes function began to arrive to work with a new model system, a simple bakers' yeast, and once again the study of a bacteriophage, this time called Mu, led to an important discovery—it was in effect a movable gene. No longer could Barbara McClintock's "jumping genes" be viewed as merely an aberration of normal gene function.

Over the next decade these pursuits resulted in a startling increase in staff and a revitalization of numerous old buildings, both laboratories and houses, along Bungtown Road.

Finding the Genes That Cause Cancer

In 1912 Peyton Rous at the Rockefeller Institute isolated from a chicken tumor a virus that when used to infect other chickens led to the formation of tumors in these chickens as well. The Rous sarcoma virus was the first of a large number of viruses found over the next several decades that were capable of inducing tumors. Initially the possibility was considered that viruses might be the cause of most cancers, but this hypothesis was soon discarded and during the next forty years tumor viruses were more a curiosity than a central focus for cancer research. This lack of interest in part reflected the general ignorance of what viruses were. They had been discovered by virtue of their disease-causing capacity and distinguished from bacteria by their very small size, but were they living? What was the significance of their DNA or RNA components? Were these components the chromosomes that carried the genetic information of these viruses? These questions were soon answered after the 1953 discovery of the double helix. Viral nucleic acid molecules are in fact tiny chromosomes carrying genes that specify not only the proteins used to construct the respective viruses, but also the enzymes needed to ensure the differential replication of the viral chromosomes in infected cells.

This understanding of the basic features of virus structure and multiplication allowed the question of how tumor viruses make cells cancerous to rise to the forefront of cancer research. In particular, the question was asked whether tumor viruses possess specific cancer-causing genes (oncogenes) on their chromosomes or do they act in an unspecific way by damaging key cellular genes involved in growth control? Key to the study of the cancer-causing process at the molecular level was the development in the early 1960s of systems in which tumor viruses transform normal cells growing in culture into their cancerous counterparts. This ability to study cancer outside of living animals allowed an enormous quickening of research. With cell cultures available for studying the replication of both DNA and RNA tumor viruses it became possible to use both genetic and biochemical approaches to identify the genes involved in virus replication as well as in their cancer-causing abilities.

A major difference between the RNA and DNA tumor viruses emerged. The cancer-causing genes of DNA tumor viruses play vital roles in the replication of these viruses. In contrast, the cancer-causing genes of RNA tumor viruses are not involved in their replication. In 1976 Michael Bishop and Harold Varmus at the University of California at San Francisco made the seminal discovery that the Rous sarcoma virus oncogene was derived from a cellular gene that had been accidentally inserted into the chromosome of the Rous sarcoma virus. Since then the cellular gene origins of many other RNA tumor virus oncogenes have been identified. In every example the original cellular genes code for proteins involved in the signaling processes that turn the commitment of cells to grow and divide on or off.

A completely different picture emerged about how the DNA tumor virus oncogenes act. They code for proteins that convert cells blocked from making DNA into cells possessing the enzymatic machinery for DNA synthesis. They do this by inhibiting proteins encoded by a special class of genes, the anti-oncogenes, that function to keep cells from making unneeded cell divisions. In humans, as well as in all other vertebrate organisms, most cells are not dividing; at any given time only a very small fraction are undergoing cell division. Largely as the result of facts generated by tumor virus research the basic signaling processes that control when cells grow and divide are now being understood at the molecular level.

Cold Spring Harbor enters the tumor virology field

Ever since Peyton Rous' 1912 discovery of a cancer-causing virus at the Rockefeller Institute the process by which a virus might cause cancer had remained a major biological mystery. Watson first learned of cancer-causing viruses as a student of Salvador Luria, and later while teaching at Harvard in 1959 he hypothesized that the chromosomes of tumor viruses contain specific cancer-causing genes (oncogenes). During the next decade Renato Dulbecco at the Salk Institute and several other scientists working elsewhere developed systems in which the cancer-causing process could be studied in cells growing in culture as opposed to in living animals and these systems opened the way for studying cancer at the molecular level.

Following Dulbecco's lead, and with the new cell culture procedures in hand, Watson believed that finding viral oncogenes had now become a practical objective if only a large enough effort could be mounted. The fact that many of the research laboratories at Cold Spring Harbor lay vacant suddenly became a real advantage since there would be ample space for the many scientists who would need to be hired for this undertaking. Federal funding for the project seemed assured since at that time congress had allocated much more money for cancer research than there were good ideas as to how to use it. Thus in 1969 Watson proceeded to hire his first tumor virologist, Joseph Sambrook, an English scientist then working on the monkey tumor virus SV40 in Dulbecco's laboratory. By midyear Watson, with Sambrook's help, had obtained a five-year grant of $1.6 million from the National Institutes of Health (NIH). Sambrook was soon joined in James Laboratory by Heiner Westphal and Carel Mulder, both coming from the Salk Institute, and by Bernhard Hirt, who had been recruited by Watson to spend a year away from his laboratory in Switzerland. The hunt for viral oncogenes was off to a fast start.

Bentley Glass galvanizes the Participating
Institutions into financial action

At the same time that the NIH grant was received the Participating Institutions of the Cold Spring Harbor Laboratory of Quantitative Biology (CSHLQB) under the leadership of their chairman Bentley Glass, the geneticist and historian of science from Stony Brook, began to provide the kind of financial support envisaged when the CSHLQB was created in 1963. The sum of $42,000 was received from the Participating Institutions in 1969, each one making a contribution of at least $3000. These unrestricted gifts from the Participating Institutions in fact represented over one third of the total donations ($124,000) to the Laboratory that year, the balance of the monies coming from major corporations and foundations.

Equally important, several additional major universities joined the list of Participating Institutions so that by 1970 the CSHLQB Board of Trustees included representatives from all of the following—Albert Einstein College of Medicine, City University of New York, Columbia University, Harvard Medical School, Long Island Biological Association, Massachusetts Institute of Technology, New York University Medical Center, Princeton University, Rockefeller University, Sloan-Kettering Institute for Cancer Research, State University of New York at Stony Brook, University of Chicago, and University of Wisconsin. At the same time the Laboratory's Board of Trustees was strengthened by the addition of several prominent members of the local community, including Robert H.P. Olney, Mrs. Alexander White, Dr. Reese Alsop, Angus McIntyre, Colton Wagner, Esq., Edward Pulling, and William Woodcock. Invaluable help was also rendered by trustee Arthur Trottenberg, then vice president for administration at the Ford Foundation. It was also at this time that the institution's name was shortened to Cold Spring Harbor Laboratory (CSHL).

James Laboratory is once again bursting at the seams

The major impact that the new higher level of federal funding for science at Cold Spring Harbor had on the Laboratory's finances was clearly evident in 1970 when it reported a total income of $1,407,865, more than double the previous year's figure of $633,235. The number of scientific and support staff had also grown dramatically since Watson had become director. Starting from a base of 47 staff members in 1968, this figure had already grown to 68 by 1969 and reached 80 in 1970. Although there was adequate bench space in James Laboratory for all of the newly hired scientists, there was no room left over for offices or a seminar room. It was obvious that an office wing needed to be added onto James Laboratory, but Watson had no idea where he would get the funds for the construction. Government funding was not an option since no federal monies were then available for capital expenses at such short notice. In an interview published in *The Long Islander,* the local Huntington, Long Island, newspaper founded by its native poet Walt Whitman, Watson optimistically expressed his hopes for an "angel" to fund this project. Miraculously this angel soon appeared in the person of John Davenport, a retired Pfizer pharmaceutical company executive whose major benefaction up to this time was the Walt Whitman Birthplace in South Huntington. When John Davenport called the Laboratory to ask whether assistance was still needed, no one knew who he was or whether he had the means to help but a visit to the Laboratory campus was arranged. During lunch with the Watsons at Osterhout Cottage, Davenport, an engineer by training, revealed that before his recent retirement he had started a tumor virus laboratory at Pfizer and that he was keen to promote such research further.

Early in 1970 Davenport transferred to the Laboratory Pfizer shares valued at $100,000, half the estimated cost of the addition to James Laboratory. Construction started in

March, with the Laboratory itself acting as the general contractor. This would not have been possible had not Jack Richards, one of the contractors approached to handle the project, indicated that he would prefer to do the job as a Laboratory employee rather than as an independent contractor. In the ensuing years Richards' continued presence on the staff as "in-house contractor" (officially superintendent, and later director, of the Buildings and Grounds Department) would allow the Laboratory to move swiftly and cost effectively to meet the changing needs of science at Cold Spring Harbor. It also meant that many aged buildings, rather than being torn down and replaced with new structures, could be saved and adapted to serve new uses in ways that benefited both science and the Laboratory's budget.

While waiting to begin work on the addition to James Laboratory one of Richards' first jobs was to oversee the installation of separate new heating plants in both Carnegie Library (the former 1905 Main Building) and the 1914 Animal House (soon to be renamed McClintock Laboratory). These two original Carnegie Institution of Washington laboratories had previously been serviced by central steam from the 1913 Power House, a system that over the years had become increasingly inefficient. Eventually the no longer needed Power House would become the headquarters for the growing staff of highly skilled craftsmen that Richards would assemble. In the spring of 1970, however, further renovations would have to wait as ground was broken for the addition to James Laboratory.

JAMES LABORATORY ANNEX
(An addition to James Laboratory)
Edelman & Salzman
1971

The office wing added to 1929 James Laboratory was designed by Harold Edelman, the architect of the newly reconstructed Osterhout Cottage (see Chapter Four). A long two-story structure hugging the steep hillside, this latest addition was sheathed in vertical board siding to match the second floor addition made to the building in 1961. Soon this sizable new wing had a name of its own, James Laboratory Annex.

The Annex was articulated as a collection of discrete units, each with its own single-slope shed roof. At the south end on the top level was the Director's Office, its roof line soaring eastward in the direction of the harbor. Between the Director's Office and James Laboratory was a series of senior scientists' offices, all on the harbor side of the corridor that led to their laboratories and enjoying distant water views. On the floor below was a seminar room featuring a long wall of glass on the harbor side with a book-lined wall opposite. At the north end of the building were secretarial offices and a flagstone-paved hall incorporating the stairs to the lower level and the passageway to James Laboratory. *(1)*

(1) James Laboratory Annex.

Completed in 1971, this addition to 1929 James Laboratory provided the tumor virologists with a seminar room (lower level) and offices (upper level) under variously sloped shed roofs.

Cold Spring Harbor joins the "War on Cancer"

With the completion of James Laboratory Annex in 1971 Cold Spring Harbor Laboratory now had the facilities to become a serious player in the field of tumor virus and cancer research. Although the continuation of this research would require ongoing support from federal grants, this was not then a source of worry. That same year President Nixon and the congress declared a "War on Cancer" and the funding of cancer-related projects became the top priority of the National Institutes of Health. The budget of the National Cancer Institute (a division of NIH) was soon doubled and most of the new monies were used to establish cancer centers such as that created at Cold Spring Harbor on January 1, 1972, which was one of the first. The budget for the new center was one million dollars per year for five years. One of the conditions imposed by the granting agency was that the Laboratory have on its support staff a professionally trained librarian. Advertisements in the local newspapers attracted the attention of Susan (Gensel) Cooper who had trained in the University of California library system. Immediately hired to fill this mandated position, she later headed the Laboratory's first Publications' Marketing Department before assuming her current role as director of the Public Affairs Department.

Neurobiology teaching begins thanks to a start-up
grant from the Sloan Foundation

Another aspect of the work at Cold Spring Harbor in the early 1970s grew out of a new line of research that many of the leading bacterial geneticists of the 1960s were now pursuing. After the genetic code had been worked out in 1966 some molecular geneticists, including Sydney Brenner, Seymour Benzer, and Gunther Stent, had begun to transfer their scientific allegiance from the bacterium *Escherichia coli* to the nervous systems of animals that they thought could eventually serve as model systems for understanding how the brain develops and functions. A new generation of scientists clearly wanted to move on from the bacterial systems that thirty years before had first delighted Max Delbrück and Salvador Luria.

Finding a home for neurobiology at Cold Spring Harbor was not difficult because there were still many largely underutilized structures on the Laboratory grounds, the legacy of the institutional reorganizations and retrenchments of the previous decade. One such building, the 1914 Animal House, had been the scene of two important discoveries in genetics by Carnegie researchers—the discovery and characterization of "jumping genes" by Barbara McClintock in the 1940s and the demonstration by Alfred Hershey in 1952 that DNA is the genetic material. In 1953 both Hershey and McClintock had moved their laboratories into the then newly completed Demerec Laboratory. Thereafter the Animal House was used mainly in the summertime by the Children's Nature Study Program. In the early 1970s the time finally came to make better use of this solidly built structure that had the necessary space for creating the teaching laboratories and lecture hall needed to launch a summer program in neurobiology.

The Laboratory approached the Sloan Foundation for the initial financial help to start the neurobiology program and late in 1970 a five-year grant of $340,000 was received which allowed the virtually abandoned Animal House to be renovated for neurobiology teaching. By the following summer the new facilities were ready and two training courses were inaugurated—a lecture course on the Basic Principles of Neurobiology and a laboratory course on Experimental Techniques that focused on the giant sea slug *Aplysia californica*, whose large nerve cells are ideal objects from which to record nerve signals electrically.

Renovation of
ANIMAL HOUSE
Renamed
McCLINTOCK LABORATORY
1971

The main renovations to the 1914 Animal House needed for neurobiology work were focused on the second floor, which was remodeled into a suite of teaching and research laboratories complete with a library/seminar room for lectures. As the first neuroscientists-in-training began walking in the footsteps of two of Cold Spring Harbor's best-known geneticists, Hershey and McClintock, in July 1971 a small dedication ceremony was held under the brick-arched portico of this former Carnegie laboratory. A simple wooden ground sign carved with the words "McClintock Laboratory" was unveiled to universal acclaim on this happy occasion celebrating both the buildings new lease on life and its historic past. It was the first of many such signs to crop up as one by one many of the older structures on the Cold Spring Harbor campus underwent renovation in the 1970s. *(2,3)*

(2) Barbara McClintock at the ceremony renaming the Animal House as McClintock Laboratory in 1971.

After the 1914 Animal House was renovated for neurobiology teaching in 1971 it was renamed McClintock Laboratory in honor of Barbara McClintock who discovered "jumping genes" here in the 1940s. At the time of its renaming the building had been "modernized" by having its brick trim painted white.

(3) McClintock Laboratory today.

The white paint has been removed from the brick detailing and the trim painted deep green to restore the building to its original appearance. (Compare with illustration *14* in Chapter Two.)

Adaptive reuse proceeds apace along Bungtown Road

While plans were being made for the renovations to the Animal House a nearly frantic search for additional living space on the grounds of the Laboratory was in progress in anticipation of the newest housing shortage that would occur when the neurobiology students arrived in 1971. It soon became clear that the most cost-effective way of gaining the necessary sleeping quarters was to turn the circa 1825 Wawepex Building into a dormitory. This was the first example of "adaptive reuse" at the Laboratory—the utilization of an old building for a new purpose in order to preserve its usefulness and architectural character for future generations. In this case the new use was quite different from the previous ones and so required a complete transformation of the building interior. At the same time the building gained heat and could thus be used not only in the summer for students in the neurobiology courses, but also year-round for the graduate students and postdoctoral fellows who were increasingly becoming a part of the Cold Spring Harbor scene.

Renovation and
Adaptive reuse of
WAWEPEX BUILDING
Edelman & Salzman
1971

Built circa 1825 as a warehouse during Cold Spring Harbor's whaling era, Wawepex Building had already been adaptively reused several times, first as a lecture hall in the 1890s, later as a supplementary summer research building, and most recently as the headquarters of the Children's Nature Study Program. In 1957 when enrollment in the Nature Study Program passed the 150-student mark, additional washrooms and an outside staircase had been installed. *(4)*

When in the 1960s the building was condemned by the Laurel Hollow building inspector the children's program moved over to the Animal House. At that time the Laboratory did not have the funds to demolish the building, and thus as the decade of the 1970s dawned Wawepex Building was still standing, just barely, at the water's edge next to Jones Laboratory. *(5)*

Then an "angel" came to its rescue. With funds contributed by Manny Delbrück, wife of Phage Course originator Max Delbrück, Wawepex Building was rescued from the brink and rehabilitated in 1971 as the sorely needed year-round dormitory for young scientists. The imaginative design from Edelman & Salzman, the architects of the newly completed James Laboratory Annex, created two floors of bedrooms clustered around a skylight-lit hall located in the center of the building. The eat-in kitchen installed in the basement had a picture-window view of Cold Spring Harbor. *(6)*

(4) Wawepex Building in the 1940s from the east.

Built as a warehouse during Cold Spring Harbor's whaling era (1836–1859), Wawepex Building (*left*) had served the Laboratory successively as a lecture hall (1890s–1920s), a research laboratory (1920s–1950s), and as headquarters of the Children's Nature Study Program (1950s–1960s). The building on the right is Jones Laboratory.

(5) Wawepex Building after conversion to a year-round dormitory in 1971.

The building was condemned by the building inspector of Laurel Hollow in the early 1960s but there was not enough money to demolish it. Later when a dormitory was needed for both young neurobiology students and tumor virologists, Wawepex Building was successfully re-adapted to meet these newest needs, gaining in the process a boiler room and kitcheonette on the basement level.

(6) Wawepex Building today.

After the building was given a facelift that included new and better windows plus a porch, the Laboratory's new Development Office was installed in the basement in 1988.

Edward Pulling gives LIBA a real mission

In the wake of the Participating Institutions effectively taking control of the Laboratory's affairs in 1963 the support from the local community, which in the summertime continued to bring its children to the Laboratory in droves for the Nature Study Program, had flagged somewhat. This soon changed after Edward Pulling became the chairman of the Long Island Biological Association (LIBA) in early 1968. Pulling's wife Lucy was the daughter of early LIBA supporter Russell Leffingwell, a banker with J.P. Morgan who rose to become president of the company and whose estate was on Yellowcote Road, just around the corner from the Laboratory in Oyster Bay. Pulling had been the founder and headmaster of Millbrook School in Millbrook, New York, and upon his retirement he and his wife moved into her late father's house and first learned of the Laboratory's need for a strong LIBA through Walter Page, then its president. Under Pullings leadership LIBA launched a $250,000 fund drive in 1972 to cover the costs of enlarging and upgrading a number of the Laboratory's facilities, including the addition of a wing along the back of James Laboratory.

The scientific staff keeps growing

The west addition to James Laboratory completed in 1972 with LIBA's help was needed because Joseph Sambrook and the other senior tumor virologists who came to Cold Spring Harbor in the early 1970s brought with them a full complement of scientific staff, including postdoctoral fellows and several graduate students, the latter mostly from the Biochemistry Department of Harvard University. Among the new senior staff were Philip Sharp from the California Institute of Technology, Michael Botchan from the University of California at Berkeley, Ulf Petterson from Uppsala, Sweden, and Walter Keller from Germany via the National Institutes of Health facility outside Washington, D.C.

During this same period Demerec Laboratory became repopulated with Hajo Delius, who arrived from Geneva, Switzerland, to set up an electron microscopy facility; Richard Roberts, who relocated from Harvard to establish a nucleic acids chemistry group; and Robert Pollack, who moved the short distance from New York University to set up a mammalian genetics group in space vacated when Alfred Hershey closed his laboratory in 1970. Pure molecular genetics was strengthened that same year by the arrival of David Zipser, who came to Cold Spring Harbor from a tenured position at Columbia University. Zipser was soon joined by Pakistani-educated Ahmad Bukhari, who had begun to work on bacteriophage Mu, a bacterial virus that causes mutations in its host *E. coli* cells. Still working in Demerec Laboratory was Ray Gesteland, who had been brought to Cold Spring Harbor by John Cairns in 1967 to study the mechanism of protein synthesis. Finally, the husband and wife team of Klaus Weber and Mary Osborn arrived on sabbatical leave from Harvard to learn how to work with tumor viruses.

Year-round housing continues to increase up and down Bungtown Road

Within five years of Watson's becoming director the total number of Laboratory employees reached 104, a more than 120% increase from the initial figure of 47 in 1968. Although the support staff was recruited mostly from the local communities, the scientific staff often arrived from quite far away and were customarily offered on-grounds housing owned by the Laboratory. This long-standing tradition at Cold Spring Harbor made sense for both practical and financial reasons—because the scientists could get to their laboratories on foot, only one family car was needed despite the Laboratory's suburban location, and most young researchers could not afford the relatively expensive housing in the immediate neighborhood. The chief reason for providing as much on-grounds housing as possible for scientific staff, however, was, and still is, the demanding nature of the science itself. First-class research is not a nine-to-five occupation. Certain types of experiments require literally around-the-clock attention, so it is important for the scientists to live close to their work.

With the need for year-round on-grounds housing growing with each new scientific appointment the adaptive reuse of Wawepex Building as a dormitory was only the beginning of a period of scouring Bungtown Road from one end to the other for similar housing possibilities. During the next twelve years practically all of the older structures comprising the Laboratory's rich and varied architectural patrimony were renovated to varying degrees (many of the laboratories sprouting "wings" in the process). The first renovation step for many of these buildings was the installation of central heating.

The 1906 Firehouse was the next building after the Animal House and Wawepex Building to be thoroughly overhauled. Already adaptively reused once—going from fire station to summer apartment house—it was remodeled again and winterized in 1972. Shortly thereafter the Laboratory was able to acquire ready-made year-round housing just a little farther north on Bungtown Road in the shape of two residential structures, now known as the Yellow House and Olney House, that predated the Bio Lab's 1890 founding. The same year that Olney House was acquired, 1973, Blackford Hall finally gained heat and thus became a year-round dining facility.

Renovation of
FIREHOUSE
Edelman & Salzman
1972

The former Cold Spring Harbor 1906 firehouse had been used for summer apartment-style housing at the Laboratory since the early 1930s when it had been towed across the harbor and then remodeled and furnished courtesy of the Women's Auxiliary of LIBA (see Chapter Three). Now in the early 1970s the interior was completely gutted and three new apartments designed by Edelman & Salzman were constructed inside the remaining late Victorian shell. *(7)*

Each apartment was equipped with an entire wall of double sliding glass doors (with protective railings) at the east end of its harbor-facing dining/living room. The topmost apartment, which incorporated the building's attic as a loft space, was first occupied by Richard Roberts and his family. Directly below lived Ulf and Brigitta Petterson, and the basement flat was occupied by Klaus Weber and Mary Osborn.

(7) Firehouse after renovation into year-round apartments in 1972.

Long in use as an apartment building at the Laboratory since its 1930 move from the village across the harbor, the Firehouse was gutted on the inside in 1972 to create three apartments for staff scientists. In this view Davenport Laboratory (today part of 1981 Delbrück and 1987 Page Laboratories) is partially visible on the right. (The Firehouse was moved once again in 1986 and repainted; see illustrations *16* and *17* in Chapter Six.)

(8) Yellow House from the harbor.

YELLOW HOUSE
(Built circa 1820)
Leased 1972
Purchased 1985

Originally built in the early nineteenth century (possibly for a sea captain but most probably for the proprietor of the neighboring Jones shipyard), this Federal-style "half-house" was privately acquired by Bio Lab director Charles Davenport in the 1930s. In 1972 the Laboratory began renting it from Dr. Davenport's daughter, Jane Davenport de Tomasi, as a residence for staff scientists. Shortly thereafter a shed-roofed dormer was installed in the harbor-facing (east) roof slope.

Although the Laboratory now owned most of the property along Bungtown Road from New York State Route 25A up to and including the Sand Spit, one of the few remaining parcels under private ownership was a small plot near the site of the former Bungtown shipyard. This waterfront property on the east side of Bungtown Road was owned by Jane Davenport de Tomasi, the daughter of longtime Laboratory director Charles Davenport and widow of Director Reginald Harris. Situated on the property was a circa 1820 Federal-style "half-house," a common type of old Long Island house so named because it had a pair of windows on just one side of the front door. Although the other "half" of the house was never built, a flat-roofed kitchen wing with a balustraded deck above had later been added to the back of the house at the water's edge.

Several newly arrived scientists, including Françoise Falcoz-Kelly and Rex Risser, immediately moved into Mrs. de Tomasi's narrow little house as soon as a rental agreement was signed in 1972. When Joseph Sambrook and his family later occupied the house in 1976 an unobtrusive shed dormer was inserted in the attic story on the harbor side. Unchanged in all other exterior respects, this putative former "captain's house" (it was probably built for a sea captain or perhaps for the manager of the Jones shipyard) has long been painted a sunny yellow color, hence the name Yellow House. The house was finally purchased in 1985 and continues to be used for staff housing today. (8,9)

190

(9) Yellow House today.

The house was finally purchased by the Laboratory in
1985 and serves today as a
single-family staff dwelling.

(10) Olney House and barn in 1911.

"View of 'North Lot' with its 62 poultry runs and two sheep and goat pastures" as pictured in the *Carnegie Institution of Washington Yearbook* for 1906. Barely visible through the trees is the circa 1885 Queen Anne-style residence that was built for Timothy Baxter Linington, a relative of the Cold Spring Harbor Joneses. The house later belonged to Manhattan dermatologist Toyohiko Campbell Takami.

OLNEY HOUSE
(Built circa 1885)
Purchased
1973

Immediately to the south of the Yellow House on the opposite side of Bungtown Road stood a house and barn that were built in the mid-1880s for Timothy B. Linington. He was a brother of the Stephen Linington who in the late nineteenth century lived with his family just down the road in Airslie. Mrs. T.B. Linington was the former Hess Lee Howard, sister of the former Katharine Seaman Howard, then wife of Townsend Jones, Jr., a nephew of Laboratory founder John D. Jones. (The misses Howard were daughters of the Reverend Robert Theus Howard, rector of St. John's Church from 1872 to 1882.)

The Linington house was later owned by Dr. Toyohiko Campbell Takami, a prominent Manhattan skin specialist who was born in Japan in 1876, the son of a samurai. With LIBA's help the Laboratory purchased the house from Dr. Takami's heirs in 1973 and subsequently renamed it in honor of Robert H.P. Olney who had led the successful LIBA fund drive that made its purchase possible and who had also served the Laboratory as treasurer before his untimely death in 1974. (10)

The house was a fine example of the Queen Anne Revival style of late Victorian architecture—very asymmetrical, with plenty of nooks and crannies, and lots of interestingly shaped windows, most of them exhibiting the distinctive Queen Anne feature of small colored panes of glass around the edges of the sashes. The house was also endowed with an elegant Doric-columned porch which wrapped around the front of the house and ended on one side in a porte-cochère, the Victorian equivalent of a carport. The sheathing material was clapboards on the ground floor and shingles above, with plain ones for the second floor and novelty cuts featured on the attic story. Originally only the ground floor clapboards had been painted, the shingles above retaining their natural appearance, but just prior to its sale to the Laboratory the entire exterior of the house had been coated with a fresh layer of white paint. *(11)*

After the Laboratory purchased the house a separate apartment was created on the ground floor, the first occupants being Michael and Ruth Botchan, and the upper floor rooms were rented to single scientists.

In the late 1970s the house needed to be repainted, but this time both the house and the barn in the back were treated to an authentic Victorian color scheme. (The barn contains the inscription "TBL 1885" which was used to help date both buildings.) The new multihued color scheme chosen to highlight all of the late Victorian architectural details was lifted from the pages of a facsimile edition (complete with color charts) of *Exterior Decoration*. This richly illustrated book written for homeowners was first published in 1885 by the Devoe Paint Manufacturing Company, one of the earliest companies to offer ready-mixed exterior house paints. *(12)*

(11) Olney House in 1973, view from the east.

The Laboratory purchased the Linington-Takami house in 1973 and afterward renamed it in honor of Robert H.P. Olney who had led the LIBA fund drive to raise the money to purchase this Bungtown Road house as a residence hall for scientists.

(12) Olney House today from the south.

To highlight its wealth of late Victorian details the house was painted in an authentic late nineteenth century color scheme as shown in *Exterior Decoration* published in 1885 by the Devoe Paint Manufacturing Company. As recommended in this book different shades were used for painting the shingles versus the clapboards.

(13) Blackford Hall after winterization in 1973.

Subsequent to gaining heat the Laboratory's reinforced concrete dormitory and dining hall had its trim painted white.

Winterization of
BLACKFORD HALL
Harold Buttrick, AIA
1973

Blackford Hall, the Laboratory's 1907 concrete dining hall and dormitory which had been built without heat for the summer Bio Lab, was the next focus of attention along Bungtown Road. The delicate task of transforming this structure into a year-round facility was successfully undertaken to the designs of Harold Buttrick, AIA, whose architectural firm had recently designed an addition to Greenvale School which was located just down Route 25A from the Laboratory. Bright blue radiators were installed in the various rooms of Blackford Hall (and the doors of the rooms were then painted bright red), but no other structural changes were needed at this time. *(13)*

The most important result of this winterization, at least as far as the Laboratory's staff was concerned, was that lunch could now be obtained on-grounds throughout the year. (This was in addition to the three meals a day that had been provided every summer since 1907.) The introduction of central heating to Blackford Hall also meant that its second floor bedrooms were now available as additional overnight accommodations for off-season visitors such as seminar speakers.

In the winterization process the basement was renovated to create a suite of offices for the Meetings Department of the Laboratory. This allowed the meetings staff to function year-round out of the same building to which scientific visitors to Cold Spring Harbor were customarily delivered by the various Long Island airport limousine services. *(14)*

(14) Blackford Hall on the eve of its 1990–1992 enlargement.

Plans for the long-needed expansion of the Laboratory's dining hall began to be implemented in the fall of 1990, with completion of the project scheduled for the spring of 1992.

*The new Tumor Virus Workshop leads to an
influential book*

Beginning in the summer of 1969 and continuing for five years the seasonal visitors
to the Laboratory included a new group, the participants in a yearly workshop created to
bring leaders in tumor virus research together with a group of "students," many of whom
were practicing scientists tempted to join the search for the putative viral oncogenes. The
first of these workshops was taught by Joseph Sambrook and London-based Lionel Crawford.
Detailed notes taken at the first workshop provided the skeleton for a textbook-style publication that presented the tumor virus field from the viewpoint of molecular biology. Many authors contributed to the final text, with Robert Pollack, James Watson, Joseph Sambrook,
Lionel Crawford, Bernard Roizman, Robin Weiss, and John Wyte writing complete chapters. After a major editing job by John Tooze, assisted by Joseph Sambrook, the 700-page
product, *The Molecular Biology of Tumor Viruses*, was published in 1973 by the Laboratory's
own Publications Department. Newly in charge of the Publications Department at this time
was Nancy Ford. Now managing editor of Cold Spring Harbor Laboratory Press (founded in
1989) she first came to the Laboratory to help with the summertime task of getting out the
annual *Symposia* volume and stayed on to oversee the editing and production of more than
20 additional titles in the Cold Spring Harbor Laboratory Monograph Series, numerous
volumes based on Cold Spring Harbor meetings and courses, and a stellar collection of
phenomenally successful laboratory manuals (an undertaking that began in 1972 with the
publication of Jeffrey Miller's *Experiments in Molecular Genetics*). After the appearance of the
tumor virus monograph in 1973 the Tumor Virus Workshop was held only once more.
With such a good text available the course had become unnecessary. In just a few years nearly 10,000 copies of the book were sold, a very substantial number considering the size of the
molecular biology community at that time.

*Whaler's Cove marina is purchased to save the inner
harbor*

Harkening back to the days when Director Demerec had held boating enthusiasts at
bay by effectively countering a dredging proposal, in 1972 the quiet beauty of summers at the
Laboratory was threatened once again by a planned major expansion of the Whaler's Cove
marina on the east shore of Cold Spring Harbor. Located directly across the inner harbor
from James and Davenport Laboratories, the 50-slip marina was built in the late 1950s on a
shore site formerly occupied by various buildings dating back to the era of the "Jones Industries." It was later purchased by Arthur Knudsen, whose marinas already dominated
Huntington Harbor. It was Knudsen's dream to expand the number of slips at Cold Spring
Harbor to 100 and to turn the adjacent historic house (built by the village miller William

(15) Pleasure craft moored in the inner harbor today.

When the owner of the Whaler's Cove marina announced plans to enlarge the number of boat slips from 50 to 100 the Laboratory successfully negotiated to buy the marina in 1973 to prevent this undesirable expansion. The marina was situated on the eastern shore of Cold Spring Harbor directly opposite the Yellow House, the Firehouse, and Davenport Laboratory (now a part of Delbrück and Page Laboratories, visible on the right in this photograph taken from the east side of the harbor).

White and shown in the circa 1885 montage in Chapter One, illustration *1*) into a clubhouse for the new enlarged marina. *(15)*

After several of the marina's immediate neighbors tried unsuccessfully to block the expansion by legal means the Laboratory's administrative director William Udry entered into negotiations to purchase the existing marina. With the help of Jerome Ambro, Supervisor of the Town of Huntington, a $300,000 purchase price was arrived at and as part of the deal Knudsen agreed to forfeit the right to open a second marina on the adjacent shorefront property that he would still own. Today the small marina continues in operation as the Whaler's Cove Yacht Club, run privately by the Whaler's Cove Association which pays a yearly rental fee to the Laboratory. Over time this rent has steadily increased and the marina remains one of the best investments the Laboratory ever made, both environmentally and financially.

Charles S. Robertson gives the Laboratory its long-needed endowment

Considering that the Laboratory was just beginning to be financially better off in the early 1970s how did it have the confidence to invest in the then seemingly high-priced Whaler's Cove marina? The story began on a day in early July of 1972 when LIBA chairman Edward Pulling excitedly telephoned Laboratory director Watson, then vacationing in California, to say that a local resident and philanthropist, Charles Robertson, proposed to visit the

Laboratory. Charles Sammis Robertson (1905–1981) had been married to the former Marie Hoffman, an heiress to the A & P Grocery Store fortune. His wife having recently passed away, Robertson was considering distributing a large portion of the assets of their family foundation, the Banbury Foundation, which had been created in the late 1950s through the sale of Mrs. Robertson's A & P shares. It was then a carefully kept secret that the Robertsons were the anonymous donors of the largest gift ever received by Princeton University (from which Charles Robertson had graduated magna cum laude in 1926). Their gift of $35 million in 1961 was used to greatly expand and generously endow the Woodrow Wilson School of Public and International Affairs at Princeton.

In the 1930s the Robertsons had assembled an estate along Banbury Lane in Lloyd Harbor that included land once owned by the family of Charles Robertson's mother, the Sammises. Because Robertson did not want these properties ever to be broken up again he hoped that Cold Spring Harbor Laboratory might find a use for his home and its adjacent outbuilding. If so, the gift would come with a proper endowment to ensure that the property never became a burden to the Laboratory.

Robertson and his legal adviser and counsel Eugene Goodwillie were given a tour of the Laboratory's attractive grounds, and afterward during lunch at Osterhout Cottage Robertson indicated that he was also considering giving Cold Spring Harbor Laboratory a separate endowment. He wisely realized that unless the Laboratory were itself financially secure it would not be able to intelligently use or lovingly care for his estate. This offer, initially too wonderful to be believed, in fact became a reality in June of 1973 when a separate corporate entity, the Robertson Research Fund, was created with a gift of eight million dollars.

Jointly managed by trustees representing both the Robertson family and the Laboratory, only the income from the Robertson Research Fund was to be used. In the early years the fund provided both research support for newly hired scientists, in advance of their soon to be forthcoming federal grants, and the monies for capital expenditures such as the renovation of badly deteriorated laboratories and the purchase of major equipment items, including an electron microscope. Under conservative management, and with a sizable fraction of the yearly income being reinvested, the Robertson Research Fund continues to provide a significant and ever-growing endowment for the scientific work at Cold Spring Harbor.

Airslie renovations bring Charles Moore Associates to the Laboratory

With the establishment of the Robertson Research Fund in 1973 the Laboratory finally had the fiscal security not only to invest in the Whaler's Cove marina to protect the inner harbor, but also to pay its new director for the first time. The salary support that Harvard University had in effect been providing for the last five years by allowing Watson to juggle his time between the two institutions would no longer be needed. Furthermore, within a year it would not be practical for the Watsons to maintain households in both Cambridge and Cold Spring Harbor as their sons would soon be in school. At the same time that plans were being made for the Watsons to live year-round at Cold Spring Harbor the possibility of their moving into Airslie arose. John Cairns, the former Laboratory director then residing at Airslie, had decided to return to England in the summer of 1973 to assume the directorship of the Imperial Cancer Research Fund Mill Hill Laboratory.

Airslie had been virtually untouched since it had been acquired in 1943 from the de Forest family to serve as the Laboratory director's residence and like many of the other Laboratory buildings was in need of renovation. What this early nineteenth century farmhouse needed most was an upgrading of its mechanical services and a restructuring of some of its interior spaces for twentieth century living. The Watsons hoped that areas used mainly for entertaining, such as the living and dining rooms, could be enlarged somewhat, perhaps by annexing additional square footage from underutilized adjacent spaces in this large rambling house. Also, a certain portion of the ground floor had to be rationalized for the realities of raising a young family. *(16)*

Among the architectural firms the Laboratory contacted concerning the renovation of Airslie was Charles Moore Associates, who at that time were just completing an innovative housing project, Whitman Village, on New York Avenue in Huntington. Charles W. Moore, a Fellow of the American Institute of Architects (FAIA) and founder of this Essex, Connecticut, firm, was then best known for several designs executed in California. His Faculty Club at the University of California at Santa Barbara made a splash with its neon supergraphics; his Kresge College on the campus of the University of California at Santa Cruz imaginatively captured the flavor of an Italian hill town; and his Sea Ranch Condominium on the California coast north of San Francisco is still celebrated today for its steep, wind-defying, multiangled shed roofs and rustic natural wood finishes, features widely emulated on both American coasts. (Coincidentally it was while staying overnight at Sea Ranch in the summer of 1972 that Watson had received the fateful telephone call from Edward Pulling reporting that Charles Robertson wished to help the Laboratory.) Currently occupying the O'Neil Ford Centennial Chair in Architecture at the University of Texas at Austin—when he's not flying off in search of challenging new architectural adventures—Charles Moore has enjoyed a long reputation not only as a designer and innovator par excellence, but also as an immensely gifted teacher and writer. He has headed up the architecture schools at the University of California at Berkeley, Yale University, and the University of California at Los Angeles and authored dozens of articles and books, including the widely quoted *The Place of Houses* (1974). His best-known recent architectural works are St. Matthew's Church in Pacific Palisades, the Hood Museum of Art at Darmouth College, and Tegel Harbor Housing in Berlin.

The Airslie renovations executed to the designs of Charles Moore Associates proved to be the beginning of an ongoing professional relationship between the Laboratory

(16) Airslie in the late 1970s.
Although this photograph was taken subsequent to the completion of the renovations in 1974, it is hard to notice any changes at first since most of them occurred inside. However, the windows and porch are new. A wine and cheese party for summer visitors is in progress on the front lawn.

and the architects from Essex. Shortly after these renovations were completed in the fall of 1974 the firm was commissioned to design a garden apartment for the nearby 1914 de Forest Stables (executed in 1975) and soon afterward a major laboratory renovation. This small group of architectural colleagues from Connecticut had by then restructured themselves as Moore Grover Harper, PC. A further name change would occur in 1983 to Centerbrook Architects and Planners. With five partners and approximately 50 designers and support staff the firm continues to operate out of the old riverside mill structure in Essex in which it was founded during Charles Moore's 1970s Yale period, designing award-winning homes (large and small) and numerous institutional projects, notably buildings for art at Dartmouth and Williams Colleges and many "houses for science" at Cold Spring Harbor Laboratory.

Renovation of
AIRSLIE
Charles Moore Associates
1974

It was at a meeting in the living room of the Watsons' Harvard-owned home at Kirkland Place in Cambridge that Charles Moore first sketched on a cocktail napkin the most memorable new feature that he was designing for the 1806 director's residence at Cold Spring Harbor Laboratory—a giant "V" flaring outward from the front door of Airslie

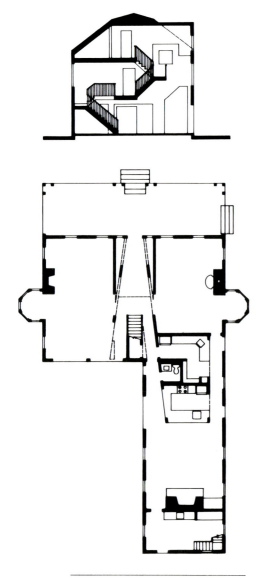

that would be superimposed on the traditional right-angled geometry of the wide Federal-style central hallway. Translated into three dimensions, the "V" was to become a soaring superstructure (with supergraphic-scale cutouts) supporting a new open staircase. Starting at the far end of the hall opposite the door the staircase would climb all the way up to the third floor of the house, doubling over itself at the top in clear view of anyone standing just inside the front door. *(17)*

Construction of this new staircase-supporting superstructure necessitated the removal of two closet-size rooms (one each on the second and third floors) directly over the front part of the hall. Their removal had the added advantage of allowing the striking topmost window at the front of the house to be seen from the ground floor. Situated in the center gable, this round-arched window was decorated on top with a rising sun motif executed by a jigsaw-toting carpenter some hundred years before. *(18)*

(17) Sectional drawing of Airslie and first floor plan dated 1974.

Renovation architect Charles Moore reinvented the center hall of Airslie with a new "V" shape that opened it up to the third floor on top and the family room at the back.

(18) Front hall of Airslie à la Charles Moore.

The new stairs were the brainchild of the celebrated architect of Sea Ranch Condominium No. 1; the top center window was the creation of an unknown nineteenth century carpenter.

Visitors to the newly renovated Airslie were delighted by being able to see all the way up to the third floor window from the front door and when descending the stairs by being able to view out of the "window" cutouts in the staircase super-structure sections of the original walls of the hallway, now brightly decorated in Colonial yellow and interestingly illuminated by architect-specified fixtures. Family and friends admired the new feeling of spaciousness (several rooms had been subtly enlarged) and the bold juxtaposition of the playful new stairs with such original time-hallowed interior features as the mellow wide-board pine floors, the Federal-style mantels with their delicately reeded columns, and the antique Franklin stoves, especially the one with the sheep, such a common sight in nineteenth century Cold Spring Harbor. Several years later the exterior of Airslie was repainted in a mid-nineteenth century color scheme to point up architectural features added after its original construction, such as the two-story bay windows at both ends of the house. *(19)*

(19) Airslie from the south lawn.

In the 1980s the house was painted in an early Victorian color scheme—Downing Cream body and Downing Earth trim—to highlight its bay windows and other nineteenth century details.

*Expansion of the summer neurobiology program leads
to a major architecture award*

Soon after the highly successful renovation of Airslie was completed the architects from Essex were called upon to breathe new life into one of the most historic buildings at Cold Spring Harbor and in fact in the scientific world. Using monies from the Robertson Research Fund the summer neurobiology program was to be expanded to include a new course on the central nervous system of mammals and an experimental workshop on how signals pass from one nerve cell to another. A laboratory facility furnished with vibration-free laboratory benches and state-of-the-art heating and ventilating systems would be needed for these purposes. At first glance the then empty late nineteenth century Jones Laboratory—the oldest marine biology laboratory building in North America still standing—might have seemed totally inappropriate for these needs, but the architects from Moore Grover Harper soon produced a highly imaginative and sensitive design solution.

Renovation of
JONES LABORATORY
Moore Grover Harper
1975

Looking more like a pristine white church than a scientific building, Jones Laboratory was the original schoolhouse and laboratory structure that the Wawepex Society erected in 1893 for the newly founded Biological Laboratory. Even the interior of the building had the look of a consecrated place when on a fine fall day in 1974 the membership of the Long Island Biological Association assembled in Jones Laboratory for their 50th annual meeting and a special anniversary address by eminent French molecular biologist Jacques Monod, a frequent summer visitor to Cold Spring Harbor. Just prior to the LIBA celebration the former maze of antique wooden cubicles and partitions had been removed from the building

in preparation for the work that was about to begin to create a new teaching laboratory for the neurosciences. For the first time in over eighty years the Laboratory's handsome beaded-board interior could be seen in all its glory. *(20,21)*

The new teaching activities would also require cubicles, but the models designed by Moore Grover Harper would bear little resemblance to the ones just removed. Sheathed in glistening aluminum, the four new cubicles were conceived as freestanding, individually climate-controlled modules, each resting on a separate concrete foundation. In an advanced neurobiology training course there is usually one instructor to three or four students and this

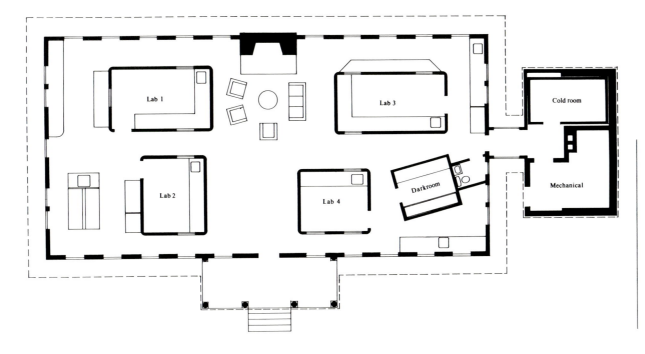

(20) Jones Laboratory today.

A small wing matching the original 1893 building was added when Jones Laboratory was remodeled in 1975 for neurobiology teaching.

(21) Jones Laboratory floor plan dated 1975.

In place of the original wooden partitions, modules clad in aluminum were inserted as "Labs 1–4." The new wing contains the "Cold" and the "Mechanical" rooms. (Compare with the original floor plan in Chapter One, illustration *13*.)

(22) Jones Laboratory's new interior.

The cubicles are new but the walls and ceiling are old. This 1975 renovation project won a coveted Honor Award for Continued Use from the American Institute of Architects in 1981.

determined the size of the new cubicles. The experiments often call for piercing individual nerve fibers, hence the necessity for isolating the modules from one another. (22)

Shared equipment and additional laboratory benches were installed along the perimeters of the modules and against the walls of the room as required. An additional module served as a photographic darkroom. That the redesigned interior looked spacious rather than cramped was due not only to the sleek design of the new modules, but also to the Laboratory's extremely tall original beaded-board ceiling which followed the contours of the high hipped roof—features clearly designed with the comfort of the first summertime marine biologists in mind.

The successful adaptation of this late nineteenth century marine biology laboratory to the requirements of late twentieth century neuroscience also required the addition of a separate wing to house the larger pieces of equipment and to provide a refrigerated storage area. Designed as a scaled-down version of the original Colonial Revival-style Jones Laboratory, the shingled addition featured a little cupola of its own on top of its miniature hipped roof.

This sensitive updating of Jones Laboratory rightly garnered numerous design awards for the architects from Connecticut, including in 1981 a prestigious Honor Award for Continued Use from the American Institute of Architects.

SV40 and the adenoviruses provide the model systems to search for cancer genes

At the same time that the neurobiology program was getting off the ground at Cold Spring Harbor it was becoming increasingly evident from the tumor virus research that the DNA animal tumor virus SV40 (a monkey virus) and the adenoviruses (similar to the common cold virus) could provide almost perfect systems for studying cancer at the molecular level, just as previously the bacteriophages and their *E. coli* bacterial host had been discovered to be a perfect system to elucidate gene structure and function. Tumor virus research at Cold Spring Harbor had initially focused exclusively on SV40, but parallel investigations soon began on the several human adenoviruses that Ulf Petterson brought from Lennart Phillipson's laboratory in Uppsala where he had received his Ph.D. By 1972 an important component of this work and in fact of most of the research at the Laboratory was the newly discovered restriction enzymes that cut DNA molecules at very specific base sequences. Carel Mulder was the first staff scientist to recognize their importance, but it was Richard Roberts working in his new facility in Demerec Laboratory who screened highly diverse groups of bacteria for new enzymes and made his reputation through their isolation and characterization. By 1974 a total of 17 new enzymes were isolated there, 11 of which showed specificities never before seen. Soon more than half of all the then known restriction enzymes had been discovered at Cold Spring Harbor and these were available for use not only by Laboratory scientists, but also by a steady stream of visitors from all over the United States and Europe. *(23)*

Restriction enzymes were used extensively by Cold Spring Harbor scientists trying to localize cancer-causing genes to specific DNA segments of tumor virus chromosomes, and in 1972 Philip Sharp, William Sugden, and Joseph Sambrook developed a powerful new way of separating on gelatin plates (gels) the DNA fragments made by restriction enzyme cutting. A year later Michael Botchan extended this method to a blotting technique developed in Scotland by Edward Southern and showed that tumor virus chromosomes are inserted into cells at fixed chromosomal locations. By 1974 the tumor virus field was in a state of great intellectual ferment. Although viral oncogenes had not yet been cleanly isolated, by then everyone knew that this breakthrough was just around the corner. Attendance at the Symposium on Tumor Viruses held that year broke all records; with 346 participants arriving to hear 115 presentations it was the most densely packed meeting yet held at the Laboratory.

A long-running and highly influential Yeast Genetics course is inaugurated

As the search for viral oncogenes was intensifying a new research initiative was inaugurated directed at the study of gene function in organisms higher than *E. coli,* specifically yeasts. Like bacteria, yeasts are single-cell organisms, but they are fundamentally different in

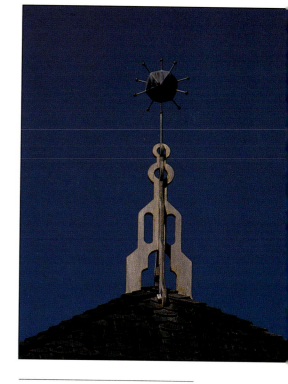

(23) Model of an adenovirus. Adenoviruses were used extensively in cancer research at the Laboratory in the 1970s. This copper model of an adenovirus is borne aloft by an anthropomorphic wooden finial. (It adorns the roof of the Gazebo erected in 1976 on top of the Laboratory's Water Treatment Plant; see also illustration *29.*)

that their chromosomes are enclosed in a membrane-bound body called the nucleus. In this sense yeast cells are much more closely related to human cells than to bacterial cells, but because they grow as single cells they are as easy to manipulate genetically as bacteria. Starting in the early 1970s an increasing number of molecular biologists began to use the bakers' yeast *Saccharomyces cerevisiae* as a model system to understand the functioning of genes of nucleated cells.

The development of the yeast frontier of molecular genetics was greatly assisted over the years by the Cold Spring Harbor training course on Yeast Genetics that was first taught in Davenport Laboratory in the summer of 1970. Its founders and longtime instructors were Fred Sherman from the University of Rochester and Gerald Fink from Cornell University. With only a one-year hiatus they continued to teach this course each summer through 1986. Meanwhile in 1973 the teaching facilities in 1926 Davenport Laboratory were modernized and the building was winterized for year-round research. In 1975 Fink and David Botstein from the Massachusetts Institute of Technology organized the first Cold Spring Harbor meeting on the Molecular Biology of Yeast, which like other course-engendered meetings at Cold Spring Harbor soon became a biennial affair continuing through 1985. By then the Laboratory could no longer accommodate all who wished to attend this highly successful comprehensive meeting and it was thereafter held elsewhere. In 1987 a smaller meeting on the subtopic of Yeast Cell Biology was held at Cold Spring Harbor and like the original yeast meeting it quickly gained a spot on the rotating schedule of meetings held in alternate years.

Distinguished academic scientists spend sabbaticals at
Cold Spring Harbor

The renovation and winterization of Davenport Laboratory in 1973 not only improved the summer teaching facilities, but also provided laboratory space for yeast geneticists on sabbatical leave from their various universities to come together to do joint research. The first such collaborative venture occurred in 1974–1975 when Botstein and Fink together with John Roth from the University of California at Berkeley spent nearly a year engaged in nonstop experimentation and conversation among themselves and with others at Cold Spring Harbor that greatly raised the level of intelligent discourse along Bungtown Road. Also in residence then was Thomas Maniatis, who was on leave from Harvard to learn how to work with animal cells in culture. Initially his visit was to be for just one year, but he stayed on an additional year when Harvard's plans for a special laboratory for recombinant DNA research were blocked by the Cambridge City Council.

This was the era in which dialogue both scientific and public often centered on whether the newly discovered recombinant DNA techniques might create new forms of life that could cause virulent new diseases or otherwise adversely affect the world's environment.

Then in March of 1975 an eminent group of scientists meeting at the Asilomar Conference Center in California called for a moratorium on experiments involving DNA from vertebrates until guidelines for recombinant DNA research could be established. How long this moratorium would last was unclear. This was a deeply disconcerting state of affairs for the scientists at Cold Spring Harbor studying oncogenes for it seemed highly likely that by using recombinant DNA procedures they might soon be able to actually isolate (clone) the cancer-causing genes of SV40 and of the adenoviruses and then determine the nature of the proteins they encoded. Laboratory director Watson thus saw the need to spend more and more of his time in the battle to come up with the guidelines for recombinant DNA research that would allow the scientists at Cold Spring Harbor and elsewhere to proceed with the research so vital for understanding cancer. In the meantime Thomas Maniatis worked on an approach to gene cloning that did not violate the moratorium. He made DNA copies of the hemoglobin RNA present in red blood cells and using the DNA sequencing techniques just developed at Harvard by Walter Gilbert and Allan Maxam he then proved that he had isolated genuine hemoglobin DNA.

Antibodies are used to localize molecules within cancer cells

Cancer cells growing in culture generally have much rounder shapes than their flattened normal counterparts. These differences reflect differences in the molecular cytoskeleton that exists in every cell. In 1974–1975 Klaus Weber and his Harvard student Elias Lazarides working in Demerec Laboratory made a great technical advance when they developed fluorescent antibodies specific for key proteins that make up the cytoskeleton. It was now possible to visualize and photograph large numbers of cells quickly and the procedure was soon widely adopted. Spurred by the new technology, a rapidly growing cell biology research group took shape on the ground floor of McClintock Laboratory.

To promote this expansion a long single-story addition with a shed roof was built along the back of McClintock Laboratory (formerly the Animal House), the same location where the original plans for this 1914 Carnegie building had called for the construction of "Future Greenhouses" (see Chapter Two, illustration *13*). Presumably those prospective greenhouses were to have been tucked into the hill that rises steeply here and this was precisely the way the laboratory addition was built in 1979, practically below grade. This modest addition was quite inconspicuous. Partially hidden by shrubbery, only its sloping roof, penetrated here and there by bubble skylights, could be seen from the back.

The new two-dimensional gel laboratory facility that was housed in the McClintock Laboratory addition allowed the cell biology group to analyze proteins on gels using computerized methods. The so-called 2-D gel technique, developed to a great extent at Cold Spring Harbor, was used to spot differences between normal and cancerous cells by comparing their protein components. The use of powerful computers to analyze the results was pioneered by James Garrels working in his QUEST (QUantitative Electrophoresis Standardized in Two Dimensions) computer laboratory which had been installed on the top floor of Carnegie Library in 1978.

Phage Mu ignites the world's interest in movable genetic elements

As cell biology research was expanding at Cold Spring Harbor in the mid-1970s a new development in basic genetics was unfolding in Demerec Laboratory. When Ahmad Bukhari arrived there in 1971 no one had anticipated the implications that would emerge from his work on the "mutator" bacteriophage Mu that he brought from Denver. As is the case with several other bacteriophages, its DNA is inserted into the chromosomal DNA of its bacterial hosts. This DNA insertion apparently occurs at random locations and it causes mutations in the gene in which the DNA settles. Understanding at the molecular level how the DNA is inserted was the problem that Bukhari and his growing research group wanted to solve. After several years the answer emerged that bacteriophage Mu was in effect a movable gene with a strong resemblance to the "jumping genes" described much earlier in maize by Barbara McClintock. At the same time at laboratories elsewhere in the United States and in Europe another class of transposable genetic elements was being described, which their discoverers called "insertion elements." Suddenly the "jumping genes" of Barbara McClintock's maize were no longer an esoteric aberration from normal genetic behavior but an important part of a major new area of research focused on the rapidly growing number of recently discovered movable genetic elements. Researchers in this new field first came together for a meeting organized by Bukhari at Cold Spring Harbor in May of 1976. Two years later an edited and updated account of work initially reported at that meeting appeared in a highly influential book published by the Laboratory under the title *DNA Insertion Elements, Plasmids, and Episomes* (1978).

LIBA helps with more housing

With the roster of scientists working at Cold Spring Harbor continually expanding there was always a need for additional housing. In 1976 LIBA once again rose to the occasion, as it had on numerous occasions since its founding over fifty years before, and launched another fund drive. Led by George J. Hossfeld, Jr., the goal of this latest campaign was to raise $225,000 for year-round housing for staff scientists. Coincidentally the funds were earmarked for a building that had figured prominently in LIBA's very first fund drive in the 1920s, the whaling era residence long known as Williams House which was situated on the former Townsend Jones land that LIBA had succeeded in purchasing for the Biological Laboratory way back in 1926.

Reconstruction of
WILLIAMS HOUSE
Moore Grover Harper
1977

Built circa 1835 as a multiple-family dwelling for workers in the Bungtown textile industry, Williams House later served the Biological Laboratory in a similar capacity for many decades starting in the late 1920s by providing apartments for the scientists who arrived with their families in the summer for research and teaching.

With a scarcity of year-round housing looming in the mid-1970s it became increasingly desirable that Williams House be winterized so that the apartments could be made available to scientists and their families in all seasons. From the start it was suspected that this would not be a simple undertaking for the building had a serious liability. It got very wet inside during rainstorms, possibly due to the plethora of natural springs that shared its hillside site, and the net result of these inundations was a cumulative legacy of decay. Thus it was not surprising when early in the design process the house was deemed useless as a substrate on which to renovate for year-round housing. The decision to erect a new residence structure was inevitable. After all of the salvageable interior fittings such as doors and mantles were removed and donated to near-

by Old Bethpage Village Restoration (Nassau County's outdoor museum of pre–Civil War life on Long Island) the structure was razed. *(24)*

(24) New Williams House from the air, looking southwest.

Williams House was reconstructed in 1977 to provide year-round housing chiefly in the form of duplex apartments. There were "lie-on-your-belly" windows in the attic of the original house (see Chapter Three, illustration *14*) where a "loft" apartment on the top floor of the main block now features the row of east-facing dormers clearly visible in this aerial photograph. The new Williams House is nearly identical in shape and size to the original.

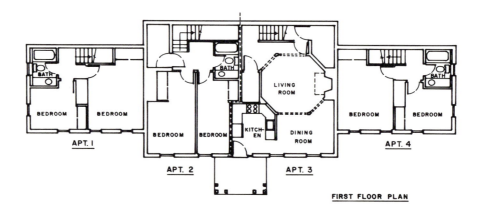

FIRST FLOOR PLAN

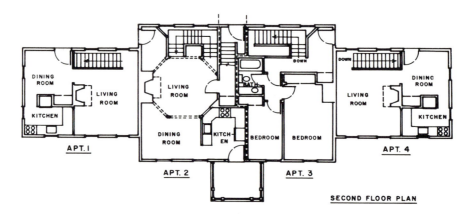

SECOND FLOOR PLAN

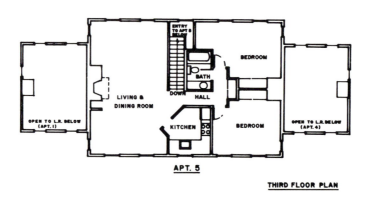

THIRD FLOOR PLAN

(25) Williams House new floor plan dated 1977.

Two of the apartments in the main block have living areas defined by octagons (note the dotted lines in the plan), more doorway than wall like the stair hall at Airslie (see illustrations *17* and *18*).

The new Williams House completed in 1977 with funds raised entirely by LIBA was almost an exact replica of the original—at least from the outside. It had the same "footprint" as the original, consisting of a main block in the center, symmetrical wings on both sides, and matching subsidiary wings at the extreme ends.

There were many differences inside, however, in part because the new structure was slightly taller than the original. Four of the new apartments in Williams

House were duplex accommodations, each occupying part of both main floors of the building, including the wings.

As designed by project architect Robert L. Harper the interiors of all of the apartments showed the imaginative use of sometimes limited space that characterized the work of Charles Moore and his associates, but the fifth apartment, situated on the top (attic) floor of the main center block, had some architectural features unique to Bungtown Road. In this top apartment the small horizontal "lie-on-your-belly" windows that were part of the design of the original Williams House were expanded into full-height dormer windows on the east-facing facade overlooking the harbor. (This device for gaining air and light was also popular with the early twentieth century modernizers of old Long Island houses.) The attic windows at the tree-shaded rear of the house, however, were still the old-fashioned "belly" kind, but also with a difference; all five were paired with state-of-the-art supplementary skylight windows placed in the roof directly above and behind them, thus preserving some privacy but ensuring light. *(25,26,27)*

(26) Interior of "loft" apartment on the third floor of Williams House.

Designed for stays of one to six months, all of the apartments in Williams House come supplied with simple furnishings.

(27) Williams House from
the back, looking northeast.

There are new ''lie-on-your-
belly'' windows at the back
of the house in the top floor
apartment, complete with
supplementary skylights.
Along this rear facade lie the
entrance doors to all five
apartments.

Where's the water to go?

Together with the expansion of year-round on-grounds housing that was already well under way by the mid-1970s, improvements and additions to general laboratory facilities were also being planned and implemented to support the greater number of researchers and the faster pace of science. These improvements included adding more offices, creating new spaces for auxiliary services, and renovating and enlarging existing laboratories. Among the basic facilities that needed to be improved were those dealing with the perennial problem of wastewater. Once again the architects from Connecticut came up with a functional and creative solution which not only solved the wastewater problem, but also enhanced the Cold Spring Harbor landscape with its first gazebo.

(28) Water Treatment Plant at the time of its completion in 1976.

This view from the east is dominated by the neo-Victorian shingled structure that incorporates the most unique Gazebo in Cold Spring Harbor on its top.

WATER TREATMENT PLANT
(With Gazebo on top)
Moore Grover Harper
1976

Ground was broken in 1975 for the construction of a Water Treatment Plant to relieve the Laboratory's dependence on leaching fields. The large, box-like, reinforced concrete plant completed in 1976 was mostly hidden from view. Because it was built into the hillside that slopes down to the harbor on the east side of Bungtown Road, the west side and the two ends of the structure were hardly visible. For the exposed east-facing side the architects had designed a superscaled portico sheathed in natural shingles accented with gingerbread trim in the corners. This soaring neo-Victorian structure facing onto the harbor at the back of the Water Treatment Plant was in fact created to function as a porte-cochère (an old-fashioned equivalent of a carport) for vehicles servicing the facility. *(28)*

(29) Gazebo from Bungtown Road.

It features a model of an adenovirus on top (see illustration *23* for a close-up view). In this view the Water Treatment Plant is completely hidden underneath the yew-bordered patio in front of the Gazebo.

Incorporated into the top of this shingled structure was a square, hip-roofed, wood-sheathed Gazebo that was reached by crossing an elegantly landscaped brick patio built right on top of the Water Treatment Plant itself and accessed from Bungtown Road. Outlined against the waters of Cold Spring Harbor the Gazebo presented a picturesque, even romantic sight which in fact has attracted several unsuspecting bridal parties who have posed here for wedding portraits. *(29)*

The Gazebo itself is topped with an anthropomorphic wooden finial in Victorian jigsaw style. The male/female figure holds aloft a tiny copper model of an adenovirus (see illustration *23* for close-up view), one of the viruses used extensively in the tumor virus research at Cold Spring Harbor.

Experiments on adenoviruses lead to the discovery of gene splicing

When the little copper model of an adenovirus was placed on top of the Gazebo in 1976 no one at the Laboratory would have predicted that the ongoing experiments with this virus would very soon revolutionize thinking about gene structure. In the spring of 1977 Cold Spring Harbor scientists made the amazing discovery that the genes of these viruses are split into physically separated regions. The primary RNA molecules made from the adenovirus DNA templates can be shortened by removal of one or more internal sections to produce much smaller RNA molecules, which in turn serve as templates for amino acids to be assembled into proteins. The removal of these RNA sections soon became known as RNA splicing. The Laboratory's experiment was done by a team that was headed by Richard Roberts and included Richard Gelinas, Thomas Broker, Louise Chow, Daniel Klessig, and James Lewis. At exactly the same time Philip Sharp, who first worked with adenoviruses as a staff member at Cold Spring Harbor, also discovered RNA splicing at the Massachusetts Institute of Technology with Susan Berget. Within weeks the phenomenon of RNA splicing was extended to SV40. Not surprisingly the 1977 Symposium on Chromatin was dominated by discussion of the implications of gene splicing.

Just as splicing was being discovered the moratorium on recombinant DNA experimentation, which had been supervised by the National Institutes of Health, was partially lifted. RNA splicing could now be looked for in systems other than viruses. Within weeks it was found in the hemoglobin gene and over the next several months it was found in virtually every vertebrate gene examined. Amazingly in all of these genes the DNA segments coding for amino acids were separated by DNA segments lacking any coding function. The discovery of RNA splicing (split genes) completely changed the way scientists thought about the genes of higher cells, and as soon as this discovery was made everyone knew that it was one of the most important events in the history of genetics.

Additional scientists soon arrived in Demerec Laboratory to exploit the RNA splicing discovery and the need for more office space became critical. With the thought that once again some structure from a bygone scientific era might be creatively reused, the Laboratory campus was scoured for underutilized space that could be turned into offices and perhaps in addition house several new departments and ancillary activities, including Graphic Arts, Safety, and a Machinist's Shop. Within a relatively short time a congenial new working environment was created for scientists and support staff alike in a building that combined old and new, but not all under the same roof.

HERSHEY BUILDING
(Incorporating a circa 1906
Potting Shed)
Moore Grover Harper
1979

The design for the new multipurpose facility was inspired by the then slowly deteriorating greenhouse complex that the Carnegie Institution of Washington had erected a few years after the completion of its 1905 Main Building (renamed Carnegie Library in 1953). Situated catercorner from the Library on the west side of lower Bung-

(30) Carnegie greenhouses circa 1930.

Originally erected circa 1906 the potting shed in the front and the greenhouse extending from the back of it were demolished in the 1940s, but the potting shed in the rear (partially obscured) plus the overall layout of the remaining greenhouses were later incorporated into Hershey Building, completed in 1979. The 1914 Animal House (now McClintock Laboratory) is partially visible on the right and 1907 Blackford Hall is in the background.

town Road, this greenhouse complex consisted of a circa 1906 potting shed plus a series of small individual greenhouses attached to it on the south side. (30)

To create the new building the old greenhouses were dismantled and then five, broad, single-story wings modeled after them in shape were erected virtually on the same foundations. The new wings varied in depth, however, from shallow at the center, where the main entrance opened onto a seminar room, to deeper on the ends. Offices for scientists were situated on one side of the seminar room and a suite of rooms for support services on the other. The two-story potting shed was preserved and

remodeled on the inside for other ancillary services, thus serving once again as a rustic wooden spine to hold together and service the attached five new wings. (31,32,33,34)

When completed in 1979 the new facility was formally dedicated as Hershey Building. In a moving ceremony three scientists with longtime Cold Spring Harbor associations paid tribute to their colleague Alfred D. Hershey, who remarked that being present at the dedication of his own building must mean that he was already on his way to "Hershey heaven," borne up there by "angels" Salvador Luria, Franklin Stahl, and Max Delbrück. (35)

HERSHEY

(31) Hershey Building, view of main entrance.

Echoing the lines of the earlier greenhouse structures, the new wings, which contain offices and auxiliary services, are attached to the former potting shed at the back.

(32) Hershey Building from the back.

The former potting shed running the full length of the building at its back dominates this view. From this angle Hershey Building looks positively ancient (at least 80 years old).

(33) Hersey Building, overall view.

The top of the potting shed can be seen along the back.

(34) Hershey Building interior.

View of the seminar area with a photograph of Alfred and Jill Hershey on the wall on the left.

(35) Alfred Hershey at the dedication of Hershey Building in 1979.

Also pictured (*left to right*) are James Watson, Franklin Stahl (University of Washington), James Ebert (Carnegie Institution of Washington), [Alfred Hershey], Max Delbrück (California Institute of Technology), and Salvador Luria (Massachusetts Institute of Technology).

Restoration work begins on the Laboratory's Victorian house on Route 25A

Just a stone's throw from the newly dedicated Hershey Building was the large dwelling that marked the entrance to the Laboratory off Route 25A. Originally built for the Fish Hatchery director, this late Victorian residence was on a downhill slide, preservation-wise. For many years it had been known as the "Carnegie dorm," and its most recent occupants, a succession of visiting scientists and their families and single scientists, could never be quite sure that the plaster ceilings would not give way. By the late 1970s the most pressing need was for a new roof, but fortunately the Laboratory's finances had improved to the point where it was possible to invest in thoroughly restoring the house, a much bigger project. In addition to installing a new roof and replacing missing Victorian trim on the exterior, the heating and wiring were overhauled, interior partitions that had been added after its initial construction were removed, and finally both the interior and exterior were redecorated to suggest what the house might have been like in the early days of occupancy by longtime Laboratory director Charles Davenport, for whom the restored house would be renamed.

That this handsome, even picturesque house was sorely in need of a new roof, not to mention a fresh coat of paint, was surely evident to anyone motoring along Route 25A. This was somewhat embarassingly confirmed in an article written by Maureen Early that appeared on January 9, 1979, in *Newsday*, the Long Island daily newspaper. Reporting about a Long Islander by the name of Roy Rasmussen who had a hobby of making dollhouses patterned after real-life houses, Early described this highly visible house on the Laboratory grounds, which just happened to be the model for Rasmussen's newest dollhouse, in the following sad-but-true terms:

> On a hillock in Cold Spring Harbor, a huge Victorian house stands in a state of obvious decay. Shriveled shingles curl on its roof; flakes of paint dangle from the white clapboard siding; the porch sags ever so slightly. Despite this, the place possesses a certain rococo grandeur—an architectural elegance of a bygone era that Roy Rasmussen felt should be preserved....

Restoration of
DAVENPORT HOUSE
1980

The restoration of 1884 Davenport House was orchestrated by the Laboratory's Buildings and Grounds superintendent Jack Richards, with the Laboratory director's wife serving in an advisory capacity. Frank Parizzi of Centerport, charged with the restoration carpentry, handcrafted on the site all of the exterior woodwork needed to

(36) Davenport House as the "Carnegie dorm" in 1978 prior to restoration.

replace parts lost through time by using sample remaining pieces as templates. On the interior he removed partitions that had been added over time to restore the original layout of the rooms. Earlier a two-bedroom apartment had been created out of most of the ground floor, including the entire kitchen wing and the original dining room and parlor. On the second floor a communal kitchen for upstairs residents had been installed in a large former bedroom, and bathrooms had been added in the centers of both the second and third floor halls. *(36)*

While the original configurations of the reception rooms on the first floor and the bedrooms on the second and third floors were being restored, plans were made for renovating the kitchen wing at the west side of the house. A new communal kitchen was installed on the ground floor of the original kitchen wing, with the handsome beaded-board dado of the original room being preserved. The second floor of the kitchen wing was substantially altered, however, with bathrooms, a laundry, and a telephone room being added.

With the exception of the separate apartment that was retained in the basement the accommodations in Davenport House were restored to what they had been originally—eight single bedrooms (four each on the top two floors of the house)—and the

(37) Davenport House today.

During restoration in 1980 the house was repainted in its original 1884 colors as determined by an historic paint color analysis.

rooms on the first floor were restored to communal use. Across the hall from the dining and living rooms was the original double parlor with its pair of marbleized slate fireplaces. This extra large reception area eventually became a Music Room for general Laboratory use, and several concert series were scheduled. *(37)*

Meanwhile an historic paint color analysis of the house had been commissioned from Frank Matero, a former classmate of Mrs. Watson's in the historic preservation program of Columbia University. Matero, who was then working in the analytical laboratories of the National Park Service's restoration division, produced an "Exterior Finishes Report" on the house, complete with hand-colored illustrations and photographs of paint layers taken through powerful Park Service microscopes. This valuable document provided all of the instructions that a painting contractor might need to restore the house to its original appearance. On the strong recommendation of Laboratory Building Committee member Elizabeth Schneider the decision was made to complete the exterior restoration by painting the house in the striking hues with which it was first decorated when built in 1884—its body painted a rich deep golden yellow, deepening to pumpkin on the gables; its trim executed mostly in deep green, but accented in certain areas with a bright yellowy green; and its window sashes painted dark maroon. In its authentic Victorian garb Davenport House became a delight to passing motorists rather than a melancholy reminder of its once proud past.

Davenport House is reborn

The story of the impending rebirth of Davenport House was written up in a *Newsday* article by Aileen Jacobson that appeared on April 19, 1979.

Laurel Hollow—Spring always brings out paint brushes on Long Island, but nowhere is it likely to bring out quite the same multi-colored bouquet of brushes as at the Cold Spring Harbor Laboratory.

The grand old Victorian house that sits on Route 25A at the entrance to the world-famous biology lab, now a shabby but sedate white, will bloom soon in a startling array of color. The colors will be the original ones: mustard yellow for the clapboard body, side shingles in darker yellow, trim in dark green, olive green and gray and windows and sashes in dark red.

The overall effect of the original was "colorful and wild but not unattractive," according to Marguerite Jones Knight [daughter of Townsend Jones, Jr.], who lived in the house as a child... .

Frank G. Matero...took 54 samples from different parts of the house, examined them under microscopes in a Boston laboratory and then subjected the bottom layers to ultraviolet lights and filters to correct for changes brought by age and lack of light.

(38) Music Room in Davenport House.

The marbleized slate mantel is original, whereas the Victorian-inspired stenciled frieze is an adaptation. All of the woodwork was "grained" in imitation of golden oak.

"The results were very positive. They match up with what we are beginning to understand about Victorian color schemes, which are foreign to our eyes today," Matero said.... The availability of ready-mixed paints after the Civil War helped contribute to Victorian color schemes.

...Matero said this may be the second exterior on Long Island, after Sagamore Hill, to be repainted using laboratory analysis.

While the house was being repainted in its rich original hues a special Davenport House Restoration Fund was established to cover the cost of redecorating the ground floor reception rooms in keeping with the exterior paint restoration. British artist Tony Greengrow used designs from Edmund V. Gillon, Jr.'s *Victorian Stencils for Design and Decoration* (1968) to create for each room a stenciled frieze that coordinated both with the warmly hued walls and with the four-foot-high dadoes of fielded paneling in each room. With the help of one of the staff scientist's teenage sons, Greengrow "grained" this paneling throughout in imitation of golden oak, a "faux finish" technique that involves first applying a light-brown base coat of paint, then using a darker shade to introduce decorative designs in imitation of the natural grain marks in wood, and finally applying varnish for protection. *(38)*

After the interior decoration of the house was completed and household furnishings purchased and installed a luncheon was held in late November 1980, with several members of the Jones and Davenport families in attendance, to officially rename the building Davenport House. To celebrate the preservation of this house as a first-class residence at the Laboratory the afternoon's guests were treated to a concert in the Music Room. Monies remaining in the privately subscribed Restoration Fund facilitated the purchase of a small grand piano for Davenport House and the presentation of several series of recitals (bassoon, flute, four hand piano, voice, etc.) between 1980 and 1982 for Laboratory staff and neighboring music lovers.

Necessary worries arise about how to enter the 1980s

While preservation plans for Davenport House were being made early in 1979 cancer research at Cold Spring Harbor also got a new lease on life when in January of that same year the National Institutes of Health changed its guidelines for recombinant DNA experimentation so as to allow work with cancer-causing genes. Within a few months many long thought about but until then forbidden experiments were done and these were reported soon afterward at the 1979 Symposium on Viral Oncogenes. It was an even bigger meeting than the 1974 Symposium on Tumor Viruses, with 141 papers appearing in the two-volume proceedings. Slowly but surely it became evident that the most interesting questions to be answered involved how the oncogenic proteins converted normal cells into their cancerous equivalents. Thus it was clear that if Cold Spring Harbor Laboratory was to remain at the forefront of cancer research, it would need to expand its facilities both for research on proteins and for the recombinant DNA procedures that had the potential to provide large amounts of those proteins that are normally present in only very small amounts. By its recent triumphs the Laboratory had in effect been the architect of its now present dilemma: If science did not continue to expand at Cold Spring Harbor, the Laboratory ran the risk of being known more for its past than its future, but just how to fund the next new growth spurt was unclear on the eve of the decade of the 1980s.

CHAPTER SIX

*The laboratories at Cold Spring
Harbor expand in all directions*

1980~1990

Before recombinant DNA was discovered, viral chromosomes provided the only systems for studying genes at the molecular level. Thus by focusing most of its resources on tumor viruses, Cold Spring Harbor Laboratory had positioned itself at the leading edge of molecular biology. As soon as it became possible to use recombinant DNA procedures, however, studying viruses was no longer the only way to do exciting science. Many other biological problems immediately became potentially solvable once ways had been found to introduce functionally active DNA into the cells of organisms other than bacteria. Such genetic engineering attempts had quickened after the discovery of recombinant DNA, and by 1977 reproducible systems had been developed for the genetic engineering of yeast as well as vertebrate cells. Shortly thereafter several young scientists who developed these procedures came to Cold Spring Harbor to exploit their systems to solve important genetic problems. As recombinant DNA techniques proliferated, major research efforts in protein chemistry, neurobiology, and mouse and plant genetics were initiated at the Laboratory, while the cell biology program was expanded to exploit several major discoveries made at Cold Spring Harbor. Increasingly with the help of corporate and foundation grants, additional facilities were created to house not only the increased staff, but also the sophisticated technologies now essential for their work.

The Key to Gene Cloning

When the genetic code was worked out in 1966 many molecular geneticists at first feared that future major advances might be very long in coming. Complete DNA molecules were far too large ever to be studied at the chemical level to determine the order of the four bases (A,G,T,C) along a given DNA chain (its sequence). Even the smaller viral DNAs contained thousands of bases and were far beyond the capabilities of even the best of chemists to tackle at the sequence level. Soon this bleak outlook changed with the discovery in 1966 by Werner Arber at the University of Geneva of a DNA cutting (restriction) enzyme that recognized specific sequences in DNA molecules. Using one of these restriction enzymes Hamilton Smith and Daniel Nathans in 1971 at Johns Hopkins Medical School cut the SV40 DNA molecule into 11 specific fragments that could easily be separated from one another. These DNA fragments were of a size that later could be sequenced by either of the two elegant new DNA sequencing methods developed in 1977 by Walter Gilbert and Allan Maxam at Harvard University and by Fred Sanger at Cambridge University. With these procedures the complete sequences of several small viral DNAs could then be determined, thereby establishing the structures of their individual genes.

At the same time a general procedure was being developed for the isolation of any DNA fragment, be it of viral or cellular origin, through the creation of hybrid (recombinant) DNA molecules. Vital to the creation of these recombinant DNA molecules was the isolation, particularly at Stanford University, of many key enzymes involved in DNA replication. Particularly important was the 1967 discovery of DNA ligase, an enzyme capable of joining together separate DNA fragments. Equally significant was the discovery that many *Escherichia coli* cells contain tiny chromosomes (plasmids) which after their isolation can be easily reintroduced into *E. coli*. Using DNA ligase Herbert Boyer at the University of California at San Francisco and Stanley Cohen at Stanford University joined a piece of vertebrate DNA to an *E. coli* plasmid. They then introduced their recombinant DNA molecule into *E. coli*. There this hybrid plasmid functioned as a tiny chromosome, reproducing every time its host bacteria cell multiplied. After this great

1973 experimental breakthrough it was clear that any piece of DNA so inserted into these tiny bacterial chromosomes could be separated away from all other DNA. Recombinant DNA thus provides a way to isolate (clone) individual genes away from all of the other DNA present in their respective cells.

Equally important now is the fabulous polymerase chain reaction (PCR) method invented in 1986 by Kary Mullis at the Cetus Corporation in California. This technique, which uses DNA synthesizing enzymes (DNA polymerases) to enzymatically copy specific DNA sequences, allows for the physical separation of any particular sequence of interest from the parental DNA molecule. Moreover, PCR has the properties of a chain reaction in that each copying cycle doubles the number of desired DNA segments, thus quickly amplifying the desired gene pieces without limit.

Soon after the wonderfully simple procedure for gene cloning was announced in 1973 fears began to be expressed that some of the novel recombinant DNA molecules might in fact be disease-causing or in some other way upset the world's ecology. If so, the great advantages to be gained from using recombinant DNA procedures to work out the structures of individual genes would be partially negated by the uncertainties as to whether the novel form of life would backfire against life on earth. By 1975 a scientist-encouraged moratorium on the use of recombinant DNA to create new forms of life came into existence, and a growing debate ensued as to whether the potential risks outweighed the potential good coming from either important new science or its use to generate commercially valuable products. Only in the fall of 1977 was the moratorium lifted for experiments on the cloning of individual vertebrate genes, with the prohibition on the cloning of potential cancer-causing genes (oncogenes) lifted only in early 1979. Since then both the science made possible by recombinant DNA and its commercial implications have far exceeded anyone's expectations. It is now virtually impossible to conceive of genetic research that does not utilize one or more recombinant DNA procedures.

How do yeasts change their sex?

One of the first genetic engineering efforts at the Laboratory focused on the common yeast, which for several years had already been the subject of both a course and a meeting at Cold Spring Harbor. Yeast cells exist as either of two sexes (mating types) called **a** and α (alpha) which under appropriate conditions fuse together. A long-puzzling observation was that a yeast cell can suddenly change from one mating type to another, but how this happened had remained a mystery. The initial breakthrough came from the University of Oregon where Ira Herskowitz and his students Jeffrey Strathern and James Hicks obtained evidence that the information to change sex exists in the form of silent DNA segments (cassettes). After the formulation of this idea Hicks went to Gerald Fink's laboratory at the Massachusetts Institute of Technology where he helped work out procedures to genetically transform yeast cells. At this point Herskowitz suggested that the place where Strathern and Hicks might go to test their cassette model using genetic engineering was Davenport Laboratory at Cold Spring Harbor, the site of the famous Phage Course in the 1940s and more recently a course on Yeast Genetics. Soon after arriving at Cold Spring Harbor Hicks and Strathern were joined by a third yeast scientist, Punjab-born Amar Klar, who came from a postdoctoral fellowship at the University of California at Berkeley where his clever genetic experiments had provided further support for the cassette model. Together they developed a yeast DNA cloning vector for isolating the DNA at the sex locus and in February of 1979 they cloned the first mating-type gene. The correctness of the cassette model was demonstrated at the molecular level not only in Davenport Laboratory, but also by the end of the summer of 1979 at the University of Washington in Seattle through similar experiments done by the English scientist Kim Nasmyth working in Benjamin Hall's laboratory.

These results were correctly seen as opening up a new field of science, not finishing one. Many more experiments would need to be done to find out how the mating cassettes moved at the molecular level. Unfortunately the yeast group had no place to work when the summer training courses were in session in Davenport Laboratory, so plans were made for an addition that would contain a large laboratory in which the yeast geneticists could continue their work uninterrupted throughout the year.

DELBRÜCK LABORATORY
(An expansion of Davenport Laboratory)
Moore Grover Harper
1981

The addition to Davenport Laboratory for yeast work was completed in 1981. The plans called for a series of offices surrounding a small, central seminar/coffee area on the top floor and a large new laboratory capable of housing up to ten scientists on the ground floor. A glassed-in passageway containing a secretarial office and a room for manipulating mating yeast cells connected the addition to Davenport Laboratory.

Similar in size and shape to the original 1926 Davenport Laboratory the new wing also had the same style—"Long Island Colonial." Its shingles wore a light stain (pale green versus the gray of the original Laboratory), its traditional-looking six-over-six windows were brightly trimmed in white, and its gable end facing onto Bungtown Road featured a semicircular attic louver at the top. *(1,2,3)*

By the time the wing was built the name "Davenport" had been reassigned to the newly restored Victorian residence at the entrance to Cold Spring Harbor Laboratory off New York State Route 25A. Davenport Laboratory, together with its new addition for yeast work, was therefore rechristened Delbrück Laboratory in honor of Max Delbrück (1906–1981), the founder of the Phage Course and mentor to an entire generation of molecular biologists. Sadly, Delbrück died less than six months before the dedication of this building in his honor.

(1) Davenport Laboratory in the 1960s.

Originally built in 1926, Davenport Laboratory later had a bridge built from it to Bungtown Road and the main entrance was changed from the ground floor of the long side facing south to the top floor of the end of the building facing the road. (Compare with illustration *5* in Chapter Three.)

(2) Davenport Laboratory interior in the 1960s.

This view shows the teaching laboratory on the top floor formerly used by the famous Phage Course (see Chapter Four) and later by the course on Bacterial Genetics.

(3) Delbrück Laboratory at the time of its completion in 1981.

It started off as a "wing" for yeast genetics added onto 1926 Davenport Laboratory (*left*), which had already been renovated and winterized in 1973. Completed in 1981, the enlarged facility consisting of the new wing and the original Davenport Laboratory was renamed in honor of Max Delbrück, the originator of the Phage Course that trained the first generation of molecular biologists at Cold Spring Harbor. In 1987 the former Davenport Laboratory gained a second "wing" (see illustrations *11* and *12*).

The first human cancer genes are cloned

As the yeast group was settling into its new quarters in Delbrück Laboratory Michael Wigler's cloning research in Demerec Laboratory, which focused on vertebrate genes, was yielding spectacular results. While a graduate student at the College of Physicians & Surgeons of Columbia University Wigler had helped develop methods for genetically engineering mammalian cells. With these methods it was now possible to look for human cancer-causing genes (oncogenes) by inserting DNA extracted from a cancer cell into a normal cell and seeing whether or not the normal cell became genetically transformed into a cancerous one. Such experiments were first successfully done in 1979 by Robert Weinberg at the Massachusetts Institute of Technology using DNA from mouse tumor donors. Weinberg later went on to show that DNA from human cancer cells could also make normal cells cancerous. Finding that an oncogene is present in a DNA sample is a very different and much easier objective than being able to separate (clone) it away from all of the other DNA in the sample. After joining the Laboratory staff in late 1978 Wigler focused on the development of recombinant DNA methods suitable for human oncogene cloning. Within a year he developed a general way to clone vertebrate genes and by 1981 he and Weinberg independently cloned an oncogene from a human bladder cancer cell, each using a different cloning procedure.

The oncogene isolated by Wigler and Weinberg turned out to be very similar to the *RAS* oncogene isolated several years earlier from a rat sarcoma cell. These oncogenes arise by mutations in the normal *RAS* genes. A single base-pair difference in the appropriate region of a *RAS* gene converts it to its oncogenic derivative. Three different human *RAS* genes exist and their oncogenic derivatives are found in a large variety of different forms of cancer. Later experiments showed that the proteins encoded by the *RAS* genes are membrane-associated molecules involved in the signal transduction pathways through which external signals initiate the chains of events that eventually lead to cell division.

Oncogenes cooperate with each other—a startling discovery

The increasing incidence of human cancer with age had long suggested that clinical cancers are the result of a number of independent mutations, each of which by itself is incapable of making a normal cell truly cancerous. Proof for this belief came from experiments in James Laboratory by Earl Ruley, who in 1984 showed that baby rat kidney cells do not become cancerous after the addition of a single oncogene, such as the *RAS* oncogene studied by Michael Wigler. Instead Ruley found that only when the adenovirus oncogene E1A was simultaneously added with the *RAS* oncogene did the rat kidney cells become cancerous. At first glance these results seemed at variance with the earlier discoveries by Wigler and Weinberg that the *RAS* oncogene alone makes a normal mouse cell line cancerous. However, in the wake of Ruley's discovery, further investigation soon revealed that the presumably normal mouse cells were in fact already on their way to becoming cancerous having previously accumulated an oncogenic mutation.

A specialized course on molecular cloning is begun

Initially very few research facilities had staff members with the expertise needed to clone genes. By 1980, however, it seemed that almost all researchers wanted to learn the techniques required to clone their favorite gene. Thus to make these techniques more commonly available the Laboratory inaugurated that same year a summer course on Molecular Cloning of Eukaryotic Genes taught by Thomas Maniatis, Edward Fritsch, and Nancy Hopkins. It quickly proved to be the most popular course ever given at Cold Spring Harbor; there were 70 applicants for its 16 places the first year and nearly 200 annually for the next several years. The laboratory protocols used in teaching the course were expanded and published in 1982 in *Molecular Cloning: A Laboratory Manual* authored by Maniatis and Fritsch together with Joseph Sambrook. By far the most successful book ever produced by Cold Spring Harbor Laboratory, it sold over 60,000 copies in its first edition alone. Because of the great popularity of the initial cloning course additional cloning courses were introduced. In 1982 Mario Capecchi and Richard Mulligan initiated a course on the Introduction of Macromolecules into Mammalian Cells. Two years later this was superseded by a new course on Advanced Techniques in Molecular Cloning, taught by Michael Smith and Mark Zoller, which featured making short DNA pieces as probes for gene isolation. This proved to be another extremely popular course.

Protein chemistry skills are necessary for recombinant DNA work

One of the most powerful applications of molecular cloning depends on the availability of advanced procedures for protein chemistry. Once the amino acid sequence of a protein is known it is theoretically possible to work backward and find the gene that codes for the protein. Many of the most interesting proteins, however, are normally present in amounts too small to be sequenced by ordinary sequencing procedures. In those few laboratories where the so-called "microsequencing" techniques were available in 1980 the scientists engaged in gene cloning had a great advantage over their colleagues working elsewhere. These new techniques were soon to be incorporated into commercially available machines, but how to pay for this advanced technology and the scientists with the expertise to utilize it presented a great dilemma as the Laboratory entered the 1980s. This kind of instrumentation was already standard equipment in the newly formed biotechnology companies such as Genentech, and in 1981 the Laboratory considered the possibility of establishing its own biotechnology company nearby. It soon became evident, however, that this was not a viable option since the only way to start such a company would be for several of the key Cold Spring Harbor scientists to move to the newly formed company and if this happened the Laboratory might lose too much of its scientific expertise.

At this point the needed financial help arrived from a most unexpected source. At a November 1981 meeting on the Patenting of Life Forms held at the Laboratory's ancillary conference facility Banbury Center (see Chapter Seven) Albert Halluin, then a patent attorney with Exxon, learned of Cold Spring Harbor's strength in DNA-related research. Exxon had already decided to inaugurate a major biotechnology division in the new laboratories they were building in Clinton, New Jersey, and they needed help in recruiting recombinant DNA experts who could find ways of using recombinant DNA-generated enzymes to transform petroleum into commercially useful products. In May of 1982 an agreement was signed initiating a joint CSHL-Exxon effort in biotechnology. No commercially secret research would be done at the Laboratory; rather its role would be to teach recombinant DNA procedures to young scientists who after spending a year or two at Cold Spring Harbor would move on to Exxon. To finance the project Exxon Research and Engineering Company would provide a grant of $1.5 million per year for five years. In return the Laboratory would make new staff appointments in the areas of protein chemistry, protein genetic engineering, and the genetics of anaerobic bacteria. Exxon was aware that the requisite research space was not then available at the Laboratory and that new facilities would need to be created. Therefore the Exxon monies for the first year would be used to construct new laboratory space.

DEMEREC SOUTH
(An addition to Demerec Laboratory)
Moore Grover Harper
1982

To house the new laboratories needed for teaching recombinant DNA techniques a two-story, oblong, aluminum-clad addition was built along the south side of 1953 Demerec Laboratory. Deep forest green in color, it blended well with the tree-studded lawn behind Davenport House next door. This "space-age bump," later called Demerec South, had two rounded chimneys constructed of the same green metal and featured a continuous band of windows on each floor. The top floor contained large

(4) Demerec South addition.

Built in 1982 to provide laboratory space for teaching recombinant DNA techniques, the addition was sheathed in dark green aluminum to blend with the surrounding foliage.

237

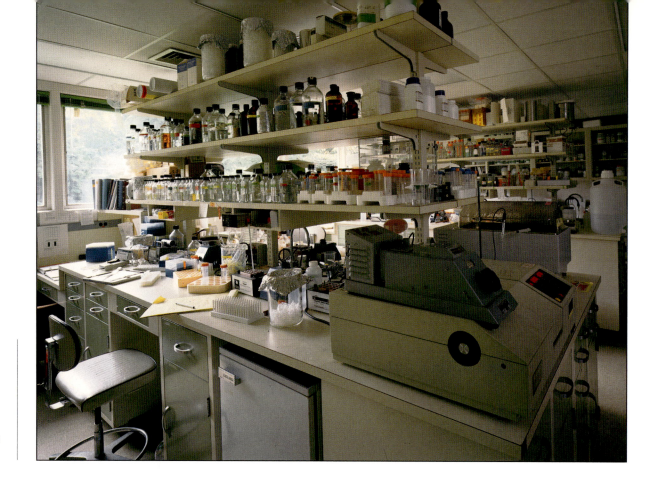

(5) Typical laboratory in Demerec South.

The wing is equipped for research in the areas of protein chemistry, protein genetic engineering, and the genetics of anaerobic bacteria.

laboratories designed for work in protein engineering and anaerobic genetics and below were laboratories equipped for the most modern techniques of protein chemistry. (4,5)

While the Demerec South addition was being built the bottom floor of Demerec Laboratory was renovated to house a laboratory kitchen facility that would serve the entire building.

The discovery of monoclonal antibodies creates the need for a new kind of facility

In 1974 George Köhler and Cesar Milstein working at Cambridge Univeristy found a way to produce so-called "monoclonal antibodies" using the spleens of immunized mice. These antibody-producing cells are the descendants of a single antibody-producing cell and thus have identical immunological specificity. In the fall of 1978 David Lane, who had already made monoclonal antibodies at the Imperial Cancer Research Fund Laboratories in London, began to make monoclonal antibodies against the large T antigen of the monkey virus SV40 in James Laboratory. Using highly purified material supplied by the biochemically inclined Robert Tjian, who had first come to Cold Spring Harbor as a junior fellow from Harvard University, Lane made a large number of these monoclonal antibodies. He later used one of them to determine that the SV40 T antigen binds tightly to a cellular p53 protein present in the nucleus.

Initially the mice needed for the production of monoclonal antibodies were supplied by the small animal facility, nicknamed the "Mouse House," that had been created in the former Sheep Shed adjacent to McClintock Laboratory to support the Laboratory's work on viral oncogenes in the late 1970s as well as to meet the needs of the cell biology group to make antibodies against cellular proteins. Previously the State University of New York at Stony Brook housed rabbits for Cold Spring Harbor scientists and also supplied them with mice and rats. Within a few years, however, the Mouse House was no longer adequate for the Laboratory's needs due to the burgeoning use of monoclonal antibodies.

At first plans were made for a combined animal house/laboratory building to be located directly across Bungtown Road from Bush Lecture Hall. Its potential cost of more than three million dollars, however, proved to be beyond the Laboratory's financial capabilities even though the National Cancer Institute had agreed to fund 75% of the originally estimated two-million-dollar cost. In addition, the proposed site did not belong to the Laboratory. It was part of a ten-acre parcel of land on the west side of Bungtown Road adjacent to Route 25A that still belonged to the Carnegie Institution of Washington. Walter Page, both a Laboratory trustee and a former member of the Carnegie Institution, was instrumental in successfully concluding negotiations in 1979 for a $200,000 purchase price for this piece of property plus the remaining ten-acre Carnegie Institution parcel adjacent to Olney House at the north end of Bungtown Road. Long Island Biological Association (LIBA) chairman Edward Pulling provided the leadership for yet another successful fund drive to purchase the property. With the land now in the Laboratory's possession and with funds to match the federal monies coming from a grant from the Max Fleischman Foundation of Nevada, plans were drawn up for an animal facility, minus the laboratory.

HARRIS BUILDING
Moore Grover Harper
1982

The new animal facility completed in 1982 was situated near the former site of the weaving factory erected by members of the Jones family. Sheathed in vertical wooden siding, the new building presented a barn-like silhouette (like its predecessor on the site) enhanced by a low shed-roofed wing at the west end. It was an innovative building for the Laboratory in that it contained Bungtown Road's first elevator installation, which was needed for transporting animal cages within the building for cleaning. *(6)*

Furnished with all of the necessities for state-of-the-art animal care and highly energy efficient, the building featured fully automated climate and light con-

(6) Harris Building.

This barn-like support facility for work using monoclonal antibodies was built in 1982 near the site of a nineteenth century textile factory that was part of the "Jones Industries."

trol. The only windows in the building were located on the ground floor where floor-to-ceiling reflective glass rounded the southeast corner of the building and flanked the lobby entrance on the east end of the structure. What looked like a long skylight window running along the roof just below the ridge was in fact an array of solar collection panels designed to supply the energy required to heat the large quantities of water needed for cleaning purposes inside the building. This new facility was named in honor of Reginald G. Harris (1898–1936), director of the Bio Lab from 1924 to 1936. At one time the residential structure known today as the Firehouse was called "Harris." This official name had never stuck, however, and it was time that an important building on Bungtown Road bore the name of Director Harris, one of Cold Spring Harbor's most charismatic scientific leaders and the originator of the annual Symposium.

Year-round neurobiology research generates exciting new monoclonal antibodies

Soon after monoclonal antibody technology was adopted for the study of oncogenes in James Laboratory it was also used to explore the specificities that underlie the ability of nerve cells to link up to the right cellular partners to form integrated nervous systems. Year-round research in neurobiology had started at Cold Spring Harbor in 1978 when Birgit Zipser set up a laboratory in the then recently renovated Jones Laboratory to study the nerve cells of the medicinal leech; she had been one of the first students in the neurobiology course that John Nicholls inaugurated at Cold Spring Harbor in 1974 using the leech as a model organism. Each nerve cord of the leech contains some 1200 nerve cells organized into approximately 30 segmentally compartmentalized groups called ganglia. Nicholls and his intellectual disciples were studying the roles played by the individual ganglia cells as well as investigating the location of the connections (synapses) that linked the nerve cells together. However, how nerve cells make their connections to each other during embryological development was still unknown.

Hoping to develop a better way of tagging nerve cells Zipser soon joined forces with Ronald McKay, who had just come to James Laboratory from Oxford University where he had tried to make monoclonal antibodies against mammalian nerve cells. Although no interesting specificities had emerged from his attempts at Oxford, when he and Zipser injected leech nerve cords into the mouse an amazingly large number of different monoclonal antibodies with many interesting specificities emerged. To help exploit these results Susan Hockfield came from the University of California at San Francisco in 1980 to use the electron microscope to localize antibody-labeled cellular molecules in the hope that some of Zipser and McKay's monoclonal antibodies might be directed against key molecules used in telling nerve cells how to link up with each other. Crucial to the inauguration of this year-round neurobiology research was the Marie H. Robertson Fund for Neurobiology created in 1977. This fund had enabled the Laboratory to offer Hockfield a staff position and late in 1980 to convene a very influential meeting on Monoclonal Antibodies to Neural Antigens.

Initially the neurobiologists in Jones Laboratory had to vacate their laboratories during the summer teaching months, but relief came when thanks to the completion of Harris Building the Mouse House became available in 1982. It was quickly converted into offices and a small laboratory for Hockfield and McKay.

Monoclonal antibody work comes to the fore thanks to several large grants

The completion of Harris Building also allowed Cold Spring Harbor to once again play an active role in mouse genetics (see Chapter Two for earlier mouse work by E. Carleton MacDowell). In 1983 a new course on the Molecular Embryology of the Mouse was taught out of Harris Building by Brigid Hogan, Frank Costantini, and Elizabeth Lacy. During the preceding several years, particularly in Frank Brinster's laboratory at the University of Pennsylvania, "microinjection" techniques had been worked out for successfully introducing DNA into the chromosomes of fertilized mouse eggs. Such genetically altered "transgenic mice" give rise to progeny mice bearing the newly introduced genes, a fact that opened up the field of mouse genetics and development to experimental manipulation.

Soon after attending the first mouse course in 1983, Douglas Hanahan, a Harvard University graduate student working at Cold Spring Harbor, began transgenic experiments in James Laboratory with funds supplied by the Laboratory. After microinjecting a hybrid gene in which tissue-specific control sequences of an insulin gene were linked to those of the SV40 oncogenic T antigen, Hanahan found T antigen expression in the transgenic offspring only in the β (beta) cells of the pancreas, the normal site of insulin synthesis. Equally important, many of the cells expressing the oncogenic T antigen gave rise to β-cell tumors.

Such experimentation, however, is expensive since running a mouse facility is itself inherently very costly. To help finance this growing program the Laboratory signed in October of 1984 a five-year, $2.1-million agreement with the Monsanto Company for a cooperative research program on the use of gene transfer in the study of gene expression during mammalian development. Under the terms of the agreement Monsanto acquired an option to develop for commercial use any inventions arising from research financed with their monies. Through this partnership the Laboratory was now in a position to try to solve one of the greatest mysteries in biology—How does an organism develop from a single undifferentiated cell into a complex system of many different types of cells, tissues, and organs? The first transgenic experiments suggested that differentiation was achieved by tissue-specific expression of genes, but many more experiments needed to be done.

By this time, however, the facilities in James Laboratory were no longer adequate, particularly those for making monoclonal antibodies, which were situated in the teaching laboratory that became unavailable during the summer months. Additional space was also needed for the growing program in mouse development to undertake new transgenic experiments. Planning for a large "James North" addition had begun in 1981, and several years later with a $400,000 grant from the Pew Charitable Trust and a $100,000 grant from the Fannie E. Rippel Foundation complementing a $425,000 grant from the National Cancer Institute the time came to start construction.

SAMBROOK LABORATORY
(An expansion of James Laboratory)
Centerbrook
1985

Completed in 1985, the major new addition on the north side of James Laboratory was named Sambrook Laboratory in honor of the researcher who first brought the techniques of tumor virology to Cold Spring Harbor—Joseph Sambrook, who was then just leaving to head the Biochemistry Department at the University of Texas Southwestern Medical Center in Dallas.

Sheathed in vertical wooden boards stained a warm gray, Sambrook Laboratory stood broad and tall in a row with similarly clad James Laboratory and Annex on the hill behind Nichols Building. The James-Sambrook Laboratory complex now totaled 17,000 square feet, the newest addition looking particularly imposing with its hip-roofed tower at the south end next to its connection to James Laboratory and its wide pavilion at the north end bearing in raised relief the Cold Spring Harbor Laboratory logo—the letters "CSH" circumscribed by a double helix. *(7,8)*

On the top floor of the building were laboratories designed for monoclonal antibody production and research, and the middle floor contained a single large research laboratory. The laboratory kitchen facilities for the entire James-Sambrook Laboratory complex were installed on the ground floor (together with the necessary mechanical equipment). All three floors were connected by an elevator, only the second such installation at Cold Spring Harbor Laboratory where most of the buildings were not so tall and the kitchens not so far removed from the laboratories.

The architects from Essex, Connecticut, who designed Sambrook Laboratory had by this time undergone a further name change. Originally Charles Moore Associates and more recently Moore Grover Harper the name of the firm had been changed to Centerbrook in 1983.

(7) Sambrook Laboratory.

Built in 1985 as an addition to 1929 James Laboratory (*left*), it contains large laboratories for tumor virology and monoclonal antibody work. The Cold Spring Harbor Laboratory logo is emblazoned on the wide pavilion on its north (*right*) end.

(8) Sambrook Laboratory from the back.

The building is sheathed in vertical board siding to match the 1961 second floor addition to James Laboratory (*right*) and the 1971 James Laboratory Annex (not shown) located on the right side of James Laboratory when viewed from the back.

Maize is again in the ascendancy at Cold Spring Harbor

The arrival of recombinant DNA procedures on the molecular biology scene which had rejuvenated mouse genetics research at Cold Spring Harbor also gave new life to the field of plant genetics. Prior to the advent of the new molecular approaches there was little interesting plant research left to be done and the field had moved into the backwater. Now it was possible not only to clone important plant genes, but also to genetically engineer plants such as the tomato and tobacco to express novel genes. In the summer of 1981 a Plant Molecular Biology course was taught for the first time in Delbrück Laboratory by the Australian John Bedbrook and Fred Ausubel from Harvard University. At the same time the scope of the work of the yeast group in Delbrück Laboratory was widened with the arrival of plant biologists Russell Malmberg from Michigan State University and Stephen Dellaporta from the University of Rhode Island. Shortly thereafter Barbara McClintock's genetic evidence for "jumping genes" (transposons) was cleanly confirmed at the molecular level just before she was awarded her Nobel prize in 1983. The time had now come to recommence experimental work on maize, and Dellaporta began focusing his attention on using transposon tagging to clone the *R* locus, long a favorite gene of maize geneticists. Essential to the Laboratory's maize efforts was the availability of fields at Uplands Farm, an 80-acre estate on nearby Lawrence Hill Road that had been donated to the Nature Conservancy upon the death of Jane Nichols, a daughter of J.P. Morgan.

The farming aspects of maize research are expensive. Fortunately an agreement signed in August of 1985 with Pioneer Hi-Bred International of Des Moines, Iowa, went a long way toward making a permanent Cold Spring Harbor maize program financially possible. The $2.5 million provided by Pioneer was to be spent over five years to support research on the use of recombinant DNA to genetically manipulate maize, the single most important crop in the United States as well as in many other parts of the world. Pioneer's large grant to the Cold Spring Harbor plant genetics effort in part reflected their recognition of the seminal role played by Carnegie scientist George Shull in the development of their company. It was Shull who in 1908 first demonstrated the improved yields of hybrid corn (see Chapter Two). Later Shull's breeding methods were first applied to the commercial production of hybrid seed by Henry D. Wallace, the editor of *Wallace's Farmer* and later secretary of agriculture and vice president of the United States, who founded the Pioneer Seed Company.

The Laboratory opens an agricultural field station nearby

Originally it was hoped that a new research laboratory for plant genetics could be built at Uplands Farm. Toward this end the Laboratory concluded an agreement with the Nature Conservancy in 1984 to purchase 12 acres, which included fields for growing maize as well as the former main garage and two of the residences at Uplands Farm. Preliminary architectural designs were drawn up for the new complex, but the project was abandoned when it became clear that the absence of any sewers along Lawrence Hill Road meant that the Laboratory would have to build a complete new sewage plant to comply with Suffolk County regulations. The decision was then made to build the new plant laboratory on Bungtown Road next to and connected to Delbrück Laboratory. Meanwhile the former garage at the Uplands Farm site was partially converted into a seed laboratory with a large storage room and its upstairs apartment was modernized for the Cold Spring Harbor Laboratory farm manager. The two existing small greenhouses at Uplands Farm were not sufficient for the expanding plant genetics program, so the Laboratory erected a new 2200-square-foot research greenhouse there in 1985. Constructed of prefabricated sections of aluminum-reinforced acrylic, the new greenhouse building was large enough to grow an entire crop of winter corn for year-round research purposes. *(9,10)*

(9) Plant genetics seed laboratory at Uplands Farm.

The seed laboratory pictured here was a 1985 adaptive reuse of the circa 1920 main garage of the former J.W.T Nichols estate on Lawrence Hill Road in Cold Spring Harbor. Known as Uplands Farm, the Nichols estate had been deeded to the Long Island Chapter of the Nature Conservancy. In 1984 the Laboratory had contracted with the Nature Conservancy to buy 12 acres for use in its plant genetics program. This purchase, which included not only fields, but also two houses and the former garage structure turned seed laboratory, was finalized in 1988.

(10) Plant genetics greenhouse at Uplands Farm.

This greenhouse of aluminum-reinforced acrylic was erected in 1985 to grow an annual crop of winter corn for experimental purposes.

A $700,000 grant from the National Science Foundation helped to cover the construction costs involved in setting up the seed laboratory and new greenhouse at the Agricultural Field Station at Uplands Farm. Pioneer allotted a $750,000 grant to help pay for the new plant genetics research laboratory on the Laboratory grounds and large contributions were also received from the Pritchard Charitable Trust, the Ira W. DeCamp Foundation, the Culpepper Foundation, and the Griggs and Burke Foundation. To provide the new laboratories and specialized facilities needed for plant biochemistry it was once again decided to add on to a preexisting structure, in this case the building first erected in 1926 as Davenport Laboratory and already enlarged once in 1981 to become Delbrück Laboratory.

PAGE LABORATORY
(An extension to Delbrück Laboratory)
Centerbrook
1987

Like the first addition in 1981, the new wing for plant genetics added to the former Davenport Laboratory, by this time re-named Delbrück Laboratory, was connected to the historic 1926 teaching laboratory core by an office-lined passageway with walls of glass. It too featured architectural details reminiscent of the "Long Island Colonial" style, including shingles stained in a pale seaworthy hue, traditional white-trimmed sash windows, and an attic louver in the west gable that faced Bungtown Road. This new wing, which was considerably larger than the first addition, was named Page Laboratory in honor of Arthur W. Page and Walter H. Page, the father and son who together with Jane Nichols Page (Mrs. Walter H.) had served on the Board of Trustees of the Laboratory and its antecedent institutions for a total of sixty unbroken years of stewardship. (11,12,13)

While Page Laboratory was under construction much of the original Daven-port Laboratory core of Delbrück Laboratory was renovated. An updated laboratory kitchen facility was installed in the basement and the top floor gained the Cummings Seminar Room directly overlooking the harbor. Named in honor of Robert L. Cummings, who served as the Laboratory's treasurer from 1977 to 1986, the Seminar Room was designed in Early American style and featured a paneled dado and ceiling beams (encasing new ones of steel) in warm-toned wood. (14)

Even with the addition of Page Laboratory in 1987 the interconnected trio of laboratories for advanced plant bio-chemistry—the "Three Ladies of Bungtown Road"—retained the small-scale, domestic-looking appearance that still characterized many of the buildings along Bungtown Road. Of the three parts of the complex Page Laboratory actually had the most square footage; its basement floor extended considerably farther north than the main

(11) Delbrück and Page Laboratories today.

Delbrück Laboratory (*left foreground*) now shares the facilities in the former 1926 Davenport Laboratory (*center*) with the more recently completed 1987 Page Laboratory (*right background*). (The sculpture in the right foreground is "Nuts and Bolts" by Michael Malpass.)

(12) Page Laboratory.

Erected in 1987 as an extension to 1981 Delbrück Laboratory, itself an extension to 1926 Davenport Laboratory, Page Laboratory houses state-of-the-art facilities for plant genetics work.

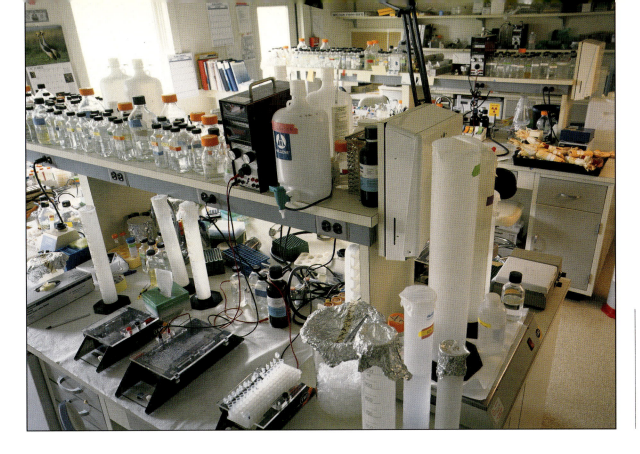

(13) Typical laboratory in Page Laboratory.

Note the labeled ears of Indian corn on the laboratory bench at the rear. Genetics work at Cold Spring Harbor once again has a maize focus.

(14) Cummings Seminar Room in the Delbrück–Page Laboratory complex.

This Early American-style Seminar Room was installed in the former Davenport Laboratory section of the complex.

block of the building, although this was cleverly hidden by a patio constructed on top of the basement extension.

Much of the enormous complement of mechanical equipment required for a building complex of this size—the total area was now 15,000 square feet—was located in the tall attic of Page Laboratory behind its bank of windows and triangular louver. The noisiest part of the air-conditioning equipment, however, was installed in its own separate building erected diagonally across the road. Partially built into the hill on the west side of Bungtown Road, this small structure was fancifully designed as an "Ice House." Its heavy concrete walls, covered with natural stained vertical boards, helped contain the noise generated by the equipment inside, while its "roof" of horizontal slats allowed for unimpeded air exchange with the outside. *(15)*

(15) "Ice House."

Built in 1987 to house much of the air-conditioning equipment for Delbrück and Page Laboratories, this structure is situated catercorner from them on the opposite side of Bungtown Road. Both distance and solid walls were used for noise abatement, while the roof was constructed of wood slats for air exchange.

(16) Firehouse about to be moved—again!

This building had to be moved once more (it crossed Cold Spring Harbor by barge back in 1930) prior to the erection of Page Laboratory on the north side of Delbrück Laboratory in 1987. Hoisted onto a temporary foundation of steel beams, it was winched along a track of greased steel rails to its new site 50 yards to the north.

The fine art of house moving is alive and well at Cold Spring Harbor

The interrelationships between the various parts of the Delbrück-Page Laboratory complex, so carefully worked out over time, make it an especially good example of how the physical plant at Cold Spring Harbor has been enhanced through sensitively designed additions to existing buildings. There is a unique footnote to this particular building expansion story, however. To make room for the construction of Page Laboratory the former Firehouse from Cold Spring Harbor, which was turned into housing for scientists after it first reached the Laboratory shores in 1930, had to be moved once again in 1986. This time the services of a professional team of building movers were needed, not those of a tugboat operator. First the Firehouse was hoisted onto a new temporary foundation of steel beams, and then, using a block and tackle, it was hauled by the movers over a specially prepared track of greased steel rails to a new harborside site 50 yards to the north. To commemorate its early history and to mark the completion of what would most probably be its final journey the building was painted a fiery red, thus providing a bright dash of color alongside the muted New England fishing village hues of Delbrück and Page Laboratories. *(16,17)*

(17) Firehouse today.

After coming to rest on its newest foundation the building was freshly painted in a color appropriate to its firehouse origin.

*Cold Spring Harbor gains a new electron microscopy
facility*

As research in plant genetics was expanding at Cold Spring Harbor in the mid-1980s studies of the cell cycle and molecular genetics in general also received a boost thanks to two powerful new technologies of modern science, electron microscopy and X-ray crystallography. Soon these became as widely used as the techniques employing restriction enzymes, two-dimensional gels, gene cloning, and monoclonal antibodies that had been pioneered in cancer-related research here.

Electron microscopy had in fact been done at Cold Spring Harbor since the mid-1950s. By the mid-1980s, however, its potential was infinitely greater as the microscope could now be connected to a powerful image-analysis computer, plus multiple monitor screens, to amplify and manipulate the observational results. How the electron microscopy staff at Cold Spring Harbor first gained its own suite of offices and laboratories in 1987 is another example of adaptive reuse of an earlier structure to serve the latest scientific needs, in this instance the tiny circa 1910 Sheep Shed that had earlier been turned into a Mouse House and then a laboratory.

CAIRNS LABORATORY
(Incorporating the
Mouse House,
An adaption in turn of a
Sheep Shed built circa 1910)
Centerbrook
1987

To provide additional space for the electron microscopy group plans were made to extend the Mouse House to the west. Originally a Sheep Shed, this bunker-like single-story structure of concrete had been used earlier as an animal facility (its Mouse House phase) and in the 1980s for preparing neurobiological monoclonal antibodies.

Completed in 1987, the new electron microscopy facility was finished in tan stucco and trimmed in green to match nearby McClintock Laboratory, the former Animal House that had been built shortly after the Sheep Shed that formed the nucleus of the new building. In the course of the enlargement the building gained a square, shingled cupola which when viewed from Blackford Hall down the steep lawn might be mistaken for a gazebo at the base of the hill. *(18,19)*

In this new facility, which was named Cairns Laboratory in honor of John Cairns, the director of the Laboratory from 1963 to 1968, cells could be magnified as

(18) Sheep Shed in 1930. Built circa 1910, it later served as an animal facility, (nicknamed the "Mouse House") starting in the late 1970s.

much as a million times. The work performed there using photographs and computer-enhanced images could provide visual answers to research questions posed both by the electron microscopists themselves and by other investigators at Cold Spring Harbor engaged in cancer and cell biology research.

An X-ray crystallography group is established

Although the appearance and behavior of cancer cells could be examined under the electron microscope, solving the mysteries of the cell cycle also demanded that the protein products of oncogenes be isolated, examined, and characterized so that the biochemical mechanisms that cause cancerous cells to continue to divide uncontrollably might be understood. If a single protein product of an oncogene could be crystallized as well as isolated, then the technique of X-ray crystallography could be used to determine the shape of that particular protein molecule and thus to understand how it physically interacts with the protein products of the other genes involved. In 1986, even before the enlarged electron microscopy facility was ready in Cairns Laboratory, an office suite for the newly hired X-ray crystallography staff was created, complete with a library/seminar room, on the second floor of Hershey Building. Although the actual laboratories for the X-ray crystallographers were located in Demerec Laboratory, the powerful computer stations designed especially for scientific applications were located in their new offices in Hershey Building.

It was partly because of the advent of these new computers that the Laboratory was able to establish a crystallography group in the mid–1980s. Prior to this neither sufficient instrument-making facilities nor the computational power needed for X-ray crystallograpy work were available at the Laboratory. By the middle of the decade, however, the situation had changed. Not only had high-powered computers become affordable and compact by then, but also the necessary rotating anode X-ray tubes and associated area detectors were

(19) Cairns Laboratory.

Incorporating the Mouse House, it was designed to house the electron micros–copy group and was com–pleted in 1987.

now commercially available, although at no modest cost. Even more important, the average time required to solve the structure of a protein of average complexity by X-ray analysis had fallen to the two-to-five-year interval associated with many other research objectives at Cold Spring Harbor.

Nevertheless, starting up a protein crystallography facility required a major financial infusion, which fortunately arrived in 1986 in the form of a grant of nearly one million dollars from the Lucille P. Markey Charitable Trust, with further help coming from the Donaldson Charitable Trust and Mr. and Mrs. Oliver Grace. Also important to the functioning of the X-ray crystallography group was a gift from the Samuel Freeman Charitable Trust that funded the installation of a modern computer facility on the lower level of the then newly completed Grace Auditorium (see Chapter Seven).

Scientists at the Laboratory discover how genes are turned on and off and that transcription factors act in pairs

The cloning and subsequent sequencing of oncogenes allowed scientists not only to determine the amino acid sequence of their respective proteins, but also to begin asking how their functioning is controlled. At any given time most of the genes of a vertebrate cell are silent; only a minority of its genes are being copied (transcribed) onto RNA templates for protein synthesis. Vital to the decision to function or not to function are regulatory proteins called transcription factors that bind to specific control (on/off-regulating) regions of genes. As the 1980s progressed more and more of the science at Cold Spring Harbor focused on the role of these transcription factors in cancer. Transcription factors turn oncogenes on or off; moreover, some oncogenes are themselves transcription factors.

An important discovery made at Cold Spring Harbor revealed that many transcription factors do not work by themselves. Instead many, if not most, genes are turned on or off only when two different transcription factors or proteins bind together prior to their attachment to the gene control region. Thus a relatively small number of different transcription factors can by pairing up in different combinations provide the needed specificity to control a much larger number of different genes. Winship Herr's work in James Laboratory on the control region of SV40 gave the first indication that in fact many transcription factors act in pairs. Soon a collaborative effort by Robert Franza, working in McClintock Laboratory, and Thomas Curran of the Roche Institute of Molecular Biology in New Jersey showed that the Jun and Fos oncogenic proteins bind to each other prior to functioning as transcription factors that turn on key genes involved in the control of cell division. Excitement about transcription factors later spilled over into the maize field when Thomas Peterson and Eric Grotewold in Page Laboratory made the exciting finding that the maize *P* control gene codes for a transcription factor whose structure is closely related to that of the Myb on-cogenic protein of vertebrates.

In addition to being a major focus of the Laboratory's research, transcription factors were also a major objective in the Protein Chemistry course started by staff scientist Daniel Marshak in the spring of 1988 in Jones Laboratory. Novel ways of purification presented in this course allowed the complete purification of a transcription factor in less than two days.

*Researchers are now going after the genes that underlie
the cell cycle*

Understanding the molecular events that underlie the ability of a cell to undergo cell division (mitosis) became a realistic objective once recombinant DNA techniques allowed the cloning of genes involved in specific stages of the cell cycle. Such genes had proved easiest to obtain in yeast, both the budding type (*Saccharomyces cerevisiae*) used by Bruce Futcher in Delbrück Laboratory and the fission yeast (*Schizosaccharomyces pombe*) used by David Beach in Demerec Laboratory. In fact, the key proteins operating during the yeast cell cycle are present in all forms of life, including maize and humans. A major feature of the cell cycle is the DNA replication process that occurs when chromosomes are duplicated just prior to cell division, and in Demerec Laboratory Bruce Stillman and his co-workers identified many of the proteins involved in SV40 DNA replication and determined their functions. Thus by the end of the 1980s the question was no longer how DNA is replicated but how its replication is regulated; in particular, how does a cell ensure that each DNA molecule is replicated only once during the cell cycle.

*Demerec Laboratory continues in an expansionary
mode*

At the time of the completion of Demerec South in 1981 it was already evident that a further addition to Demerec Laboratory would be needed. Although some of the scientific staff had their offices in Hershey Building, there was only limited space available in Demerec Laboratory for the increased secretarial help and the growing number of postdoctoral fellows and graduate students. With the expansion of the cell biology effort following quickly on the heels of the recent installation of the X-ray crystallography group in Demerec Laboratory the need for more space became critical. A seminar room was also needed for meetings of the individual research groups. Finally in 1989 construction was begun on the long-anticipated second addition to Demerec Laboratory.

(20) North side of Demerec Laboratory at the time of its completion in 1953.

This is where the Demerec North addition was later erected. Demerec Laboratory is on the right and Bush Lecture Hall on the left.

DEMEREC NORTH
(An addition to Demerec Laboratory)
Centerbrook
1989

The second addition to 1953 Demerec Laboratory took the form of a shallow single-story wing erected on the building's north side to contain offices and a seminar room. Just as the previous addition which faced south onto a grove of evergreens was compatibly sheathed in dark green metal, the architects from Centerbrook chose grayish brown brick for the exterior sheathing material of this new wing which faced north directly along the rear wall of Bush Lecture Hall to tie it in visually with both the concrete of Demerec Laboratory and the brick of the Lecture Hall. These two buildings were now in such close proximity that the small area of turf remaining between them was replaced with an equally compatible flower- and shrub-bordered terrace composed of large gray flagstones with feature squares of brick. *(20,21)*

A shallow curved canopy of lead-coated copper, echoing that over the main entrance to Demerec Laboratory, faced onto the terrace and marked the entrance to this second addition, soon known as Demerec North.

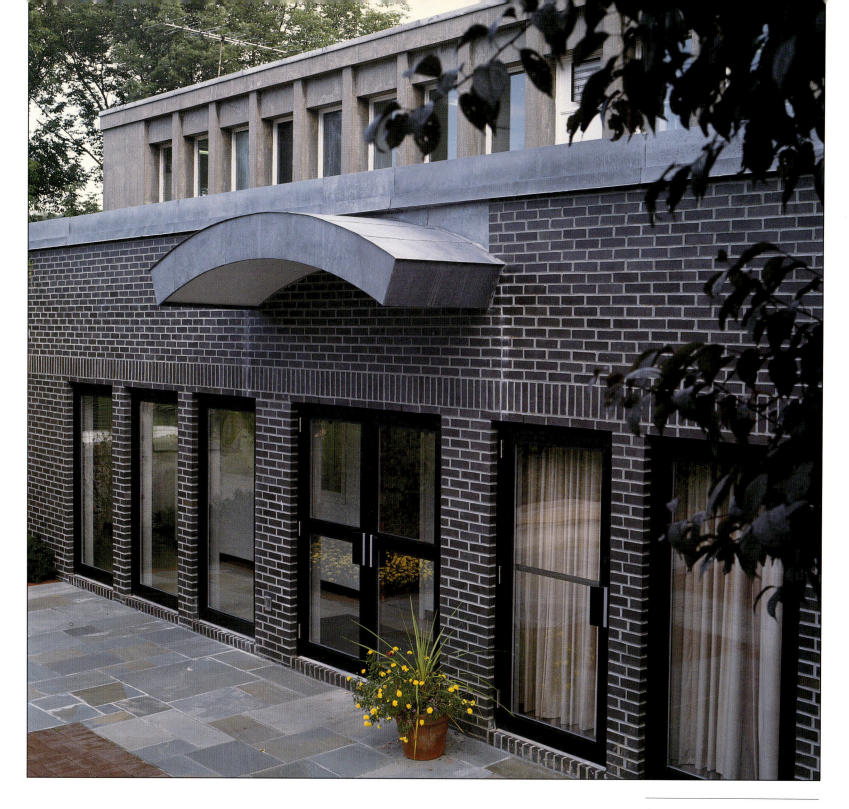

(21) Demerec North addition.

Completed in 1989, it houses offices and a seminar room for cell biologists working in Demerec Laboratory.

Monoclonal antibodies help explain how DNA viral oncogenes work

As the Demerec North addition neared completion the long-established tumor virus research in James Laboratory yielded an extremely important discovery about how viral oncogenes cause cancer. In 1983 the London-trained American Edward Harlow had joined the James Laboratory group and proceeded to attempt to use monoclonal antibodies to find out how the adenovirus oncogenic protein E1A makes cells cancerous. He sought to determine which cellular proteins coprecipitate (bind) with the E1A protein after addition of monoclonal antibodies specific for E1A and, if he were lucky, he would discover that one or more of the cellular proteins that bound strongly to the E1A protein functions in the control of cell growth and division.

Working in the Sambrook Laboratory monoclonal antibody facility and assisted by a grant from Amersham International, Harlow's research group observed that at least five specific proteins bind tightly to the E1A protein. Initially he had no idea as to the functional identity of these cellular proteins. Then in 1988 Peter Whyte, a postdoctoral fellow in Harlow's laboratory, read a paper in *Nature*, the British science weekly, describing the molecular properties of the RB (retinoblastoma) protein. It was the first-discovered member of a new class of proteins which function to prevent a cell's progression through the cell cycle. The genes that code for these growth inhibitory proteins are known as antioncogenes. The size of the RB protein was very similar, if not identical, to one of Harlow's five E1A-binding proteins, specifically the 106K protein. A collaboration was quickly initiated with scientists in Robert Weinberg's laboratory at the Massachusetts Institute of Technology and it was soon discovered that the 106K protein and the RB protein were in fact identical.

This very major discovery suggested that the E1A protein makes cells cancerous by binding to and thereby neutralizing several cellular proteins that normally function to prevent unwanted cell divisions. Strongly backing this idea were subsequent experiments performed at Cold Spring Harbor Laboratory, the Massachusetts Institute of Technology, and the National Institutes of Health laboratories in Bethesda, Maryland, which revealed that the oncogenic proteins encoded by SV40, polyomavirus, and papillomavirus also bind to the 106K protein and to several other proteins now known to be anti-oncogenic as well.

After a twenty-year search by Cold Spring Harbor scientists the general mechanism by which DNA tumor viruses make cells cancerous was at last known. Their oncogenic proteins neutralize the key cellular proteins that regulate the cell cycle, and thus uncontrolled DNA replication is allowed to occur. A year later, in 1989, this observation was extended through the finding by Harlow and Robert Franza, in McClintock Laboratory, that a second E1A-binding protein, known as 60K, associates with a key cell cycle control protein, called CDC2, which was being studied in David Beach's laboratory in Demerec Laboratory.

From the gene to genome

In 1986 several Laboratory scientists became involved in the debate over whether the time had come to work out the complete genetic instructions of human beings. One afternoon session of the annual Symposium, which that year focused on the Molecular Biology of *Homo sapiens,* was devoted to the pros and cons of the proposed Human Genome Project. Because three billion base pairs would need to be sequenced many in the audience, particularly those in the younger age groups, opposed the project, being worried that if the endeavor was not controlled it could effectively bankrupt modern biology. In contrast a number of the more veteran scientists, including the Laboratory's director, favored going ahead, believing that it would immeasurably advance the course of future medical practice. They argued for the project to start soon, provided it emphasized technological improvements and would encompass the sequencing of the genomes of more simple model organisms, including bacteria, yeast, and *Drosophila.*

Over the following two years the United States congress was persuaded to provide federal funds to commence the Human Genome Project, with the year 2005 the target date for its completion. To help make this program a reality James Watson accepted the offer of James Wyngaarden, then the director of the National Institutes of Health, to become the director of the National Center for Human Genome Research. Watson knew that assuming this position while at the same time remaining director of the Laboratory was bound to become one job more than he could effectively handle over the long term. Thus to help maintain the future continuity of research at the Laboratory Bruce Stillman was appointed to the position of assistant director in the summer of 1990.

Several diverse activities marked the Laboratory's initial involvement with the Human Genome Project. An annual meeting on Genome Mapping and Sequencing was inaugurated in 1988 and a course on the Cloning and Analysis of Large DNA Molecules was organized by Nat Sternberg of du Pont in 1989. The following year Thomas Marr, an informatics expert from the Los Alamos National Laboratory, moved to the Laboratory to work with Richard Roberts on computer-assisted genome analysis and to assist David Beach with his project to map and hopefully eventually sequence the 15-million-base-pair genome of the fission yeast *Schizosaccharomyces pombe.*

*A major increase in federal funding for science must be
achieved before the end of this century*

In retrospect the 1980s were equally as successful for science at Cold Spring Harbor—if not more so—as the 1970s had been, and this was accomplished despite an ever-worsening situation for science funding in the United States. The arrival of the recombinant

Working Out the Complete Genetic Instructions for Human Existence

As this century draws to a close the world of genetics is expanding its horizon from an interest in the gene per se to the task of unraveling the complete genetic instructions (genome) for several key forms of life. The amount of genetic information carried by the genomes of even the simplest organisms is staggering and the task of determining all of the genes of say the bacterium *Escherichia coli* is far from trivial. To do so the exact order of some five million bases (A,G,T,C) must be established, a multiyear sequencing project some 20-fold larger than any yet accomplished. Despite the magnitude of the task there is great enthusiasm for rapidly moving toward this goal. Once we have established the exact structure of the *E. coli* genome, conceivably by 1995, we will automatically know the number and types of proteins that it specifies and have for the first time a description of all the protein machinery underlying the functioning of a complete cell.

There also exist plans for the total sequencing over the next decade of the genomes of the intensely studied roundworm *Caenorhabditis elegans* and of the geneticist's long-favored tool the fruit fly *Drosophila melanogaster*. Both these genomes are some 20-fold larger than that of *E. coli* and carry the information for some 10–20 thousand genes. When their genomes are sequenced we shall possess two complete programs for the development of fertilized eggs into sexually mature adult organisms. All of the information for the correct turning on and off of the switches that control selective gene functioning at last will be on hand, giving those biologists who study embryological development powerful tools to eventually totally describe the molecular events underlying the differentiation of unspecialized cells into complex appendages such as wings or legs, each of which is made up of a large number of different cell types.

The ultimate genome to be sequenced is that of ourselves. Essentially all of our genetic information is found on our 22 different autosomes and the X and Y sex chromosomes. On average, the single DNA molecule that comprises each of our chromosomes contains some 100 million base pairs, roughly the same number found in the entire *Drosophila* genome. The task of sequencing all of the human DNA is indeed a herculean one and has been called biology's "moon shot." Through this mammoth sequencing endeavor we will eventually be able to identify all of the human genes, a number that will most certainly exceed 100,000. The key uncertainty is how soon the cost of sequencing can be reduced to less than a dollar a base pair so that the entire project can be completed for a cost of no more than three billion dollars.

The Human Genome Project need not be completely finished before it yields real payoffs for biology and medicine. Already the genetic markers and gene probes it has generated have begun to greatly advance the finding of genes responsible for important genetic diseases. For example, the gene responsible for the fragile-X syndrome, a major cause of mental retardation, has just been isolated, as well as several genes that lead to predispositions to important human cancers. With this information at hand genetic diagnostic procedures will soon be available to those families who so want to use them.

There remain, however, many important disease genes still to be isolated. For example, both schizophrenia and manic-depressive disease have genetic components, as does early-onset Alzheimer's disease. Once we discover the genes underlying these diseases we will know the exact chemical defects that cause them and will be better able to target research in those directions most likely to yield eventual cures.

To make use of the knowledge generated by the Human Genome Project the peoples of this world must become much more genetically literate, not only understanding Mendel's laws, but also how the four letters of the DNA alphabet (A,G,T,C) are used to encode genetic information. The gene as well as the atom will be indelible features of the human landscape for as long as human civilizations prevail and for them to be dealt with wisely they must be broadly understood by all layers and divisions of human society.

DNA era had meant that greater sums of money were needed to exploit the vast potentialities of the new techniques. Yet federal funding only kept pace with inflation and had proved inadequate to fund all the superior science that then could be done. In fact, if supplementary funding from Exxon, Monsanto, Pioneer, and Amersham had not flowed into Cold Spring Harbor the history of science here in the 1980s would have been very different. The Laboratory would have had neither the resources to attract and hold the scientists necessary for the major role it played in the recombinant DNA revolution nor the funds to provide the facilities and sophisticated technologies needed for their work. However, as the first decade of recombinant DNA research ended, all of the major pharmaceutical and biotechnology companies had their own DNA experts and no longer needed to rely on academic partners. Thus the answer to the question of whether Cold Spring Harbor Laboratory and other major centers of biology could continue to live up to their potential in basic research was bound to depend on whether there would be a major increase in federal funding in the last decade of the twentieth century.

CHAPTER SEVEN

Cold Spring Harbor Laboratory enters its second century as a university of DNA

The Cold Spring Harbor Laboratory that commenced its second century in 1990 presented a very different facade from that which existed at its founding. It had taken on the appearance of a mini-university as opposed to that of a summer camp for biologists. By then the complexity of modern DNA research demanded more than could be provided in the existing historic buildings along Bungtown Road. To keep pace with the expanding scope of molecular biology and to accommodate the increasing numbers of scientists engaged in unraveling its mysteries a large new auditorium was built near the entrance to the Laboratory and shortly thereafter an upper campus site was developed to allow the extension of its research objectives into neurobiology.

Long a familiar name in international scientific circles through its meetings, courses, and books, the Laboratory gained increased national recognition through two outreach programs established at nearby off-campus sites—the Banbury Conference Center in Lloyd Harbor and the DNA Learning Center in Cold Spring Harbor—programs firmly rooted in the educational tradition of the original Biological Laboratory founded in 1890.

*Poster for 1990
Cold Spring
Harbor
Symposium
on Quantitative
Biology*

Cold Spring Harbor Laboratory
55th Symposium on Quantitative Biology

Organized by:

Eric R. Kandel, *Columbia University College
of Physicians & Surgeons*
Terrence J. Sejnowski, *The Salk Institute*
Charles F. Stevens, *The Salk Institute*
James D. Watson, *Cold Spring Harbor
Laboratory*

Registration materials may be obtained from:

Meetings Coordinator
Cold Spring Harbor Laboratory
Cold Spring Harbor, NY 11724
(516) 367-8346

*The fifty-fifth symposium will focus on
the following topics:*

*Signal Transduction in the Nervous System
Neural Plasticity and Learning
Development
Sensory Systems
Motor Systems
Cognitive Neuroscience
Computational Neuroscience*

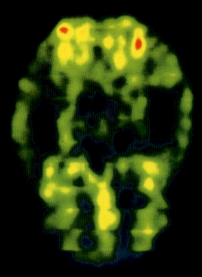

The Brain
May 30 - June 6, 1990

Speakers include:

Richard W. Aldrich, *Stanford University School of Medicine*
Per Andersen, *University of Oslo*
Richard A. Andersen, *Massachusetts Institute of Technology*
David J. Anderson, *California Institute of Technology*
Richard Axel, *Columbia University College of P&S*
Jay M. Baraban, *The Johns Hopkins University School of Medicine*
Denis A. Baylor, *Stanford University School of Medicine*
Howard C. Berg, *Harvard University Biological Laboratories*
Emilio Bizzi, *Massachusetts Institute of Technology*
Colin Blakemore, *University of Oxford*
Friedrich Bonhoeffer, *Max-Planck-Institut*
Tobias Bonhoeffer, *The Rockefeller University*
James M. Bower, *California Institute of Technology*
John H. Byrne, *University of Texas Health Science Center, Houston*
Constance Cepko, *Harvard Medical School*
Jean-Pierre Changeux, *Institut Pasteur*
Martha Constantine-Paton, *Yale University*
William Maxwell Cowan, *Howard Hughes Medical Institute, Bethesda*
Francis H. C. Crick, *The Salk Institute*
Thomas Curran, *Roche Institute of Molecular Biology*
Antonio R. Damasio, *The University of Iowa*
Robert Desimone, *National Institute of Mental Health*
Gerald M. Edelman, *The Neurosciences Institute*
Gerald D. Fischbach, *Washington University School of Medicine*
Apostolos P. Georgopoulos, *The Johns Hopkins University School of Medicine*
Claude Ghez, *Research Foundation for Mental Hygiene, Inc.*
Charles Gilbert, *The Rockefeller University*
Michael E. Goldberg, *National Eye Institute*
Patricia Goldman-Rakic, *Yale University School of Medicine*
Corey S. Goodman, *University of California, Berkeley*
Paul Greengard, *The Rockefeller University*
Sten Grillner, *Karolinska Institutet*
Stephen Heinemann, *The Salk Institute*

Neville Hogan, *Massachusetts Institute of Technology*
John J. Hopfield, *California Institute of Technology*
H. Robert Horvitz, *Massachusetts Institute of Technology*
David H. Hubel, *Harvard Medical School*
A. James Hudspeth, *University of Texas Southwestern Medical Center, Dallas*
Julian J. B. Jack, *The Physiological Society, Oxford*
Lily Y. Jan, *University of California, San Francisco*
Yuh N. Jan, *University of California, San Francisco*
Thomas M. Jessell, *Columbia University*
Ken Johnson, *The Johns Hopkins University School of Medicine*
Edward G. Jones, *University of California, Irvine*
Bela Julesz, *Rutgers University & California Institute of Technology*
Eric R. Kandel, *Columbia University College of P&S*
Stanley B. Kater, *Colorado State University*
Mary B. Kennedy, *California Institute of Technology*
Christof Koch, *California Institute of Technology*
Mark Konishi, *California Institute of Technology*
Stephen G. Lisberger, *University of California, San Francisco*
Margaret Livingstone, *Harvard Medical School*
Rodolfo Llinas, *New York University Medical Center*
Gerald E. Loeb, *Queen's University, Kingston, Canada*
U. J. McMahan, *Stanford University School of Medicine*
Brenda Milner, *Montreal Neurological Institute & Hospital*
Mortimer Mishkin, *National Institute of Mental Health*
Richard G. M. Morris, *University of Edinburgh Medical School*
Vernon B. Mountcastle, *The Johns Hopkins University School of Medicine*
J. Anthony Movshon, *New York University*
Ken Nakayama, *The Smith-Kettlewell Eye Research Institute*
William T. Newsome, *Stanford University School of Medicine*
Roger A. Nicoll, *University of California, San Francisco*
Shosaku Numa, *Kyoto University Faculty of Medicine*
Illustration courtesy of Semir Zeki, University College London

Dennis D. M. O'Leary, *Washington University School of Medicine*
Paul H. Patterson, *California Institute of Technology*
Gian F. Poggio, *The Johns Hopkins University School of Medicine*
Tomaso Poggio, *MIT Artificial Intelligence Laboratory*
Dale Purves, *Washington University School of Medicine*
William Quinn, *Massachusetts Institute of Technology*
Marcus Raichle, *Washington University School of Medicine*
Pasco Rakic, *Yale University School of Medicine*
Louis F. Reichardt, *University of California, San Francisco*
Werner Reichardt, *Max-Planck-Institut fur Biologische Kybernetik*
David A. Robinson, *The Johns Hopkins University School of Medicine*
Edmund T. Rolls, *University of Oxford*
Gerald M. Rubin, *University of California, Berkeley*
Bert Sakmann, *Max-Planck-Institut fur Medizinische Forschung*
Richard H. Scheller, *Stanford University*
James H. Schwartz, *Columbia University College of P&S*
Peter H. Seeburg, *Universitat Heidelberg ZMBH*
Terrence J. Sejnowski, *The Salk Institute*
Carla J. Shatz, *Stanford University School of Medicine*
Wolf Singer, *Max-Planck-Institut*
David L. Sparks, *University of Pennsylvania*
Larry R. Squire, *University of California, San Diego*
Charles F. Stevens, *The Salk Institute*
Michael P. Stryker, *University of California, San Francisco*
Masatoshi Takeichi, *Kyoto University*
Vincent Torre, *Universita di Genova*
Richard W. Tsien, *Stanford University Beckman Center*
Shimon Ullman, *MIT Artificial Intelligence Laboratory*
David Van Essen, *California Institute of Technology*
Gerald Westheimer, *University of California, Berkeley*
Torsten N. Wiesel, *The Rockefeller University*
Paul F. Worley, *The Johns Hopkins University School of Medicine*
Robert H. Wurtz, *National Eye Institute*
S. M. Zeki, *University College London*

*Professional meetings at the Laboratory are now held in
an award-winning auditorium facility*

As attendance at several of the Laboratory's most popular summer meetings regularly began to exceed 400 participants in the late 1970s it was obvious to all that 250-seat Bush Lecture Hall was no longer an adequate facility to sustain a thriving meetings program. For years the overflow crowd had "attended" these meetings via closed-circuit television installed in the basement of Blackford Hall and in the seminar rooms of McClintock and James Laboratories, but this was not a satisfactory solution when the time came for the question and answer period that followed each presentation. Finally in the fall of 1984 excavation began for a much larger auditorium. The site chosen for the new facility was just to the east of 1982 Harris Building (with its commodious parking lot) and thus on the opposite side of Bungtown Road from the Laboratory's 1907 reinforced concrete dining hall cum dormitory Blackford Hall (the birthplace of the annual Symposium). The fact that this location was almost entirely a steeply sloped hill did not deter the architects from Essex, Connecticut; in fact it suggested certain environmentally friendly features that could be incorporated into the design.

Although a small part of this hill had been excavated just after the turn of the century for the sand to make the concrete for Blackford Hall, a large amount of earth remained to be carted away before construction could begin on the new auditorium. There was also a great deal of money to be raised. This building project was the largest yet undertaken by the Laboratory and would require a massive infusion of new funds. The Long Island Biological Association (LIBA) once again rose to the challenge, with its chairman Edward Pulling mounting his fourth fund drive. This time the initial goal was two million dollars. As completed, however, the final cost of the building was four million dollars since the originally conceived one-story building had grown to include a full basement floor equipped with a large computer facility, the end result being a building of more than 15,000 square feet.

GRACE AUDITORIUM
Centerbrook
1986

Firmly planted perpendicular to Bungtown Road and just 100 yards north of its intersection with New York State Route 25A, brick-clad, slate-roofed Grace Auditorium is today the first stop for visitors arriving at the Laboratory. Clearly visible from the highway, the portico of the building features on its fascia shiny golden lettering honoring Oliver and Lorraine Grace of Oyster Bay and New York City for their generous donation to the ambitious fund-raising campaign that made this facility a reality. *(1)*

This broad and deep south-facing portico with its stout columns denotes the

(1) Grace Auditorium.

Completed in 1986, Grace Auditorium is the newest venue for the annual Cold Spring Harbor Symposium on Quantitative Biology. The year after it was built it won an ARCHI award from the Long Island Chapter of the American Institute of Architects. Its striking dormer windows are visible from Route 25A.

front entrance to the building at the northern edge of a large parking area that parallels the highway. At the time of the completion of Harris Building in 1982 this parking area had been attractively landscaped with pin oaks, white pines, and willows, as well as highway-concealing low evergreen shrubs and a rustic, stream-spanning bridge. The landscape design by Innocenti & Webel was executed under the supervison of Laboratory grounds foreman Hans (Buck) Trede.

Behind the entrance portico the front doors open onto the south end of a spacious lobby flooded by light from a tall, deep dormer window above the entrance. The auditorium space extends the full depth of the building on the west (left) side of the lobby. Just inside the front door on the east (right) side are the offices of the Public Affairs Department, and farther north on the same side of the lobby, bathed in light from two skylights in the shallow hipped roof, is the staircase to the lower level where additional Laboratory departments are housed. These include the Meetings Office, which sets up shop upstairs in the lobby during meeting registrations; the QUEST Laboratory (*QU*antitative *E*lectrophoresis *S*tandardized in *T*wo Dimensions), which was relocated from the top floor of Carnegie Library; and the Samuel Freeman Computer Center. The Laboratory's Book Store, which offers popular nonfiction titles and standard texts as well as essential sundries for visitors, is also located on the basement level. The lower level offices receive natural illumination from deep light wells positioned along the perimeter of the building.

Designed in 1983 by Centerbrook (the successor firm to Moore Grover Harper), Grace Auditorium was completed in 1986. The builders were A.D. Herman Construction Company of Huntington Station. Recipient of a 1987 ARCHI Design Award from the Long Island Chapter of the American Institute of Architects, Grace Auditorium owes its special quality to its eclectic style. From 1884 Davenport House across the road it borrows late nineteenth century architectural motifs and colors, echoing the dormer window theme (also popular elsewhere along Bungtown Road) of that Victorian residence as well as its palette of deep rich trim colors. The architects also adopted many classical design concepts and details from the laboratories built at Bungtown around the turn of the century, for example the Auditorium's entrance portico and the covered loggia that extends along its principal south-facing facade. There is even a row of quasi-dentils, fabricated of short lengths of copper piping, below the roof cornice.

By far the most striking feature of Grace Auditorium is the array of dormer windows along the top of the building facing due south toward Route 25A. One of these dormers is oriented in such a way that winter sunlight falls directly onto two heat-storing Trombe walls of masonry that form part of the south side of the 360-seat auditorium space. These interior walls are painted a mauve-tinged slate gray which sets them off from the blues and greens that dominate the color scheme of the auditorium space. This restful palette is highlighted just below the ceiling by a con-

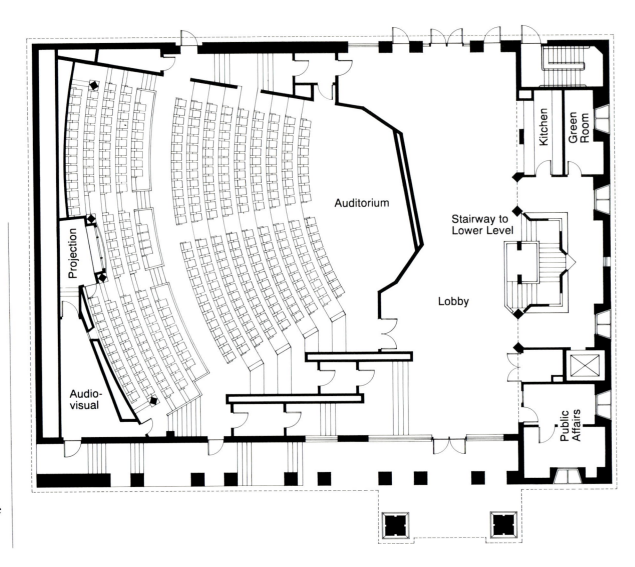

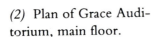

(2) Plan of Grace Auditorium, main floor.

Angling the auditorium space created alternative entrance possibilities for latecomers or browsers who can thus slip in or out of the talks unobtrusively, a feature especially appreciated during the Symposium. The two parallel, truncated, interior walls on the left side of the auditorium (*bottom*) are Trombe walls of solid masonry which store solar heat. During the winter the sun's rays directly strike these walls through one of the tall dormer windows on the roof of the Auditorium.

tinuous deep slate blue freize bearing a computer-generated double helical DNA design. In addition to the passive solar heating achieved by the Trombe walls, a passive energy-saving device for cooling was designed into the building by situating the auditorium space on the west side of the building which is partially below grade. *(2,3,4)*

The lobby space, generously lit at both ends by broad, glazed front (south) and rear (north) doors and floor-to-ceiling windows, is large enough for coffee breaks to be held indoors in inclement weather. Ordinarily during break periods the meeting participants congregate on the deep patio situated directly outside the north-facing lobby doors just across the road from Blackford Hall.

The patio is enclosed on the west and north by the tall hill that was first partially excavated prior to the construction of

(3) Auditorium space of Grace Auditorium.

With a 360-seat capacity and superb acoustics (note the wooden baffles in the ceiling), it has become a favorite venue for community meetings and concerts. The decorative dark blue frieze along the top of the wall is a computer-generated image of a DNA molecule.

(4) Grace Auditorium lobby, looking south toward the main entrance.

Visible in the right foreground are the stairs to the lower level where the Meetings Office and the Samuel Freeman Computer Center are located.

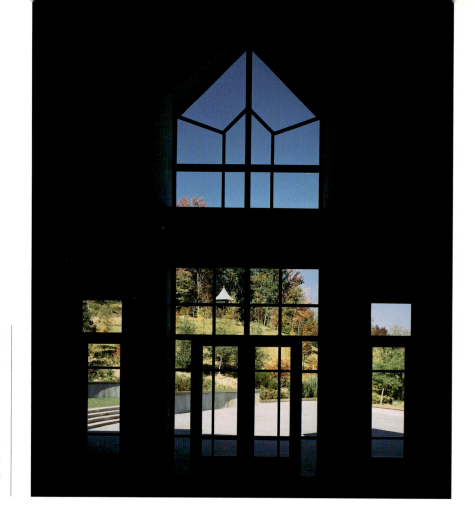

(5) Grace Auditorium lobby, looking north toward the rear entrance.

Just outside the doors on the extreme left is the base of the curving terraced walkway up to the Octagon (the copper-roofed gazebo structure in the distance near the center of the photograph).

Blackford Hall in 1907. When a more substantial portion of the hill was removed to make room for the erection of Grace Auditorium, both the appearance and stability of the hill had suffered. Fortunately this apparent liability could be turned into an asset by clever designing. On its west exposure the hillside was assiduously stepped into miniterraces that end in a curving pathway to the top of the hill where a pretty gazebo beckons walkers. This focal point of the Grace Auditorium landscape design affords a panoramic view encompassing some of the oldest structures at the Laboratory and a vista clear across the harbor at its head—at least in the winter. Topped with a flaring octagonal roof of copper oxidized to a pale aqua color (which used to sit atop the pumphouse of the former Henry C. Taylor estate on Snake Hill Road in Cold Spring Harbor), the gazebo is appropriately called the Octagon. The roof is supported by a circle of columns, interconnected with Victorian-style brackets, which was designed by the architects from Centerbrook and painted in shades of green highlighted with rose to match the trim colors of Grace Auditorium down below. *(5, 6)*

Although the west side of the hill leading up to the Octagon was now terraced, the hillside to the north remained to be stabilized. This was accomplished by the planting of dozens of trees, specifically native types such as ashes and oaks, from the top of the hill all the way down to Bung-

(6) Octagon.

The roof of the Octagon came from a 1920s pumphouse which was located on the former Henry C. Taylor estate in Cold Spring Harbor. It was relocated to the Laboratory grounds and later placed atop the eye-catching new columns in 1986.

town Road. To further ensure that the slope would eventually knit back together the ground was seeded with wild grasses. In time as the canopy of native trees filled out these grasses would die and the area would take on the appearance of a native woods. Ornamental grasses, shrubs, and flowering trees were planted along the edges of the patio and the terraced area to provide a variety of color and interest throughout the year against the backdrop of forest trees. Designed by Keith Simpson Associates of New Canaan, Connecticut, the comprehensive landscape plan for Grace Auditorium not only solved the potential erosion problems of the site, but also made the flagstone-paved patio next to the building an inviting place to "talk science."

Guest accommodations at the Laboratory are expanded thanks to former visitors

The completion of Grace Auditorium in 1986 meant that the larger meetings of 200 to 400 participants could now be more comfortably accommodated on Bungtown Road, at least as far as the formal presentations and poster sessions were concerned (the latter now staged in the newly vacated Bush Lecture Hall). However, this also meant that there were increasing numbers of visitors to be housed, not to mention fed, preferably on the Laboratory grounds. There were fewer than 150 beds available on-grounds in 1986 for meeting participants, and many of these were set up in an ad hoc fashion in the summer apartments that were vacant prior to the arrival of the summer course instructors and their families for whom these seasonal accommodations were designed. As a result of this on-site housing shortage many visitors had to spend nearly an hour each day in transit to and from area motels. The first step toward providing additional housing accommodations for meeting attendees was the completion in 1989 of six new guest cottages, which were quickly dubbed the "New Cabins." Designed in log-cabin style, the New Cabins echoed the feeling of the first "guest cottages" created on the Laboratory grounds in 1951 when the single-room wooden cabins on the lawn between Blackford Hall and McClintock Laboratory were dismantled and then combined on the hill on the west side of Bungtown Road as Cabins A, B, C, and D. To fund construction of the New Cabins the Laboratory turned to those who would most appreciate the need for improved guest housing on the Laboratory grounds. For the first time past participants in Cold Spring Harbor Laboratory's numerous courses and meetings were solicited for gifts. The response was very gratifying. Donations were received from over 150 grateful alumni, including very generous gifts from Herbert Boyer, founder and chairman of Genentech, and Thomas Maniatis and Mark Ptashne, cofounders of Genetics Institute, Inc.

NEW CABINS
Centerbrook
1989

Erected on a hilly wooded site west and north of Grace Auditorium, the six New Cabins completed in 1989 combined features of both the old 1951 Cabins (of which two still remain today) and the late 1950s units of Page Motel (demolished in 1987). Designed by the architects from Centerbrook to utilize prefabricated log cabin building technology (with all building parts furnished precut and ready for assembly), each of the small new "H"-shaped cottages contains four double bedrooms and two baths. In each cabin two bedrooms are lined up back to back (with a bath in between) on both sides of a tall center section that is architecturally distinguished by a double-height cathedral ceiling. The inside surfaces of the sturdy log walls were pickled to lighten the interiors and even out the wood tones of the knotty pine. Named in honor of a group of living scientists with strong ties to Cold Spring Harbor—Herbert Boyer, Harry Eagle, Bentley Glass, Salvador Luria, Franklin Stahl, and Gunther Stent—the new guest cottages accommodate a total of 48 seasonal visitors. (7,8)

With the season for courses and meetings at Cold Spring Harbor already extending further into early spring and even early autumn in the late 1980s, provision was made for heating the New Cabins as the season requires. The New Cabins were such a hit that plans were made for an additional five to be built on a high, gently sloping wooded site to the west of the first six. One of these latest cabins (which were completed in 1991) would be named Alumni Cabin.

(7) New Cabins.

Completed in 1989, the six New Cabins each sleep eight visitors. They were built of a single thickness of logs plus other building parts prefabricated off site. The logs were stained on the exterior and pickled on the interior.

(8) Typical New Cabin bedroom.

Visible through the window on the right is Hazen Tower, part of the Neuroscience Center completed in 1991. Note the framed photograph of tents on the lawn of Blackford Hall, an earlier version of summer visitor housing at Cold Spring Harbor.

Plans are made to enlarge Blackford Hall to keep pace
with the Laboratory's expanding programs

Although the visitor housing situation was greatly improved by the construction of the six New Cabins (and the five additional ones soon to be completed), as the Laboratory's centennial year drew to a close in 1990 an even more important design for the increased comfort of summer visitors was on the drawing boards, the enlargement of 1907 Blackford Hall. After Grace Auditorium was completed in 1986 the lunch lines in the Laboratory's dining hall sometimes extended not only through the dining room itself, but also out and across Bungtown Road. Green and white striped tents for overflow dining and serving quickly became a summertime fixture on the lawn behind Blackford Hall. Even before the completion of Grace Auditorium an addition to the dining hall had been contemplated. In 1988 plans were drawn up for a $3.5 million enlargement of Blackford Hall that would increase the dining capacity, improve service, and upgrade the kitchen facilities.

As the time to begin construction of the Blackford Hall addition drew near, a tactical decision was reached to approach corporations in the pharmaceutical and bio-technology industries both in the United States and abroad for financial assistance. The rationale was that these companies benefit importantly from the existence of this unique non-profit educational institution on the western shore of Cold Spring Harbor. Scientists from more than 200 such companies had attended the Laboratory's courses and meetings, and the books, laboratory manuals, and journals published by the Laboratory were now an important resource in their research and development laboratories. Companies responding early to the Laboratory's bid for help included Boehringer Mannheim, Bristol Myers Squibb, Burroughs Wellcome, Du Pont, Hoffmann-LaRoche, Merck, Sharp & Dohme, Pall Corporation, Rorer, Schering-Plough, SmithKline Beecham, and Westvaco Corporation.

To minimize disruption of summer visitor services the construction of the Blackford Hall addition was designed to be executed in two stages over the course of two winters. The first stage of the project, inaugurated in the fall of 1990, involved excavating a sub-basement below the original building and laying adjacent, but not touching, foundations immediately to the north and east for an enlarged kitchen wing and a two-story extension to the dining room. Besides a recreation room and additional storage space in the new sub-basement, the enhanced facilities in Blackford Hall would comprise a new, high-ceilinged garden dining room at grade level, directly accessible (together with its adjacent bathrooms and bar) to visitors congregated on the sloping rear lawn. On the main floor above this would be two interconnecting dining rooms, the original and its extension to the east, the latter opening onto a new alfresco dining deck. In total the dining capacity of Blackford Hall would be increased from 170 to more than 400 per seating, and as a result of the enlargement of the food preparation area the kitchen would now be able to serve nearly 50% more meals per day. When completed in the spring of 1992 the new Blackford Hall would be more than double the size of the original building, increasing from roughly 10,000 to over 22,000 square feet.

The Laboratory's advanced courses and professional meetings have expanded into spring and fall

Typically the Laboratory's visitor season opens with a two-week course session that begins in early April. This is followed by a five- or six-week-long series of professional meetings, one per week, that ends with the Symposium on Quantitative Biology, Cold Spring Harbor's first and longest-running annual meeting which was begun in 1933. The end of the Symposium signifies the beginning of the "high season" of postgraduate training courses, now organized in the main as three three-week sessions with three to four courses each session offered primarily for students at the Ph.D. level and beyond; in fact, university professors and department chairman, including Nobel prizewinners, have been students in courses at Cold Spring Harbor together with researchers from industry. Another series of professional meetings then takes place, including the long-running annual Phage Meeting (now incorporated in the Molecular Genetics of Bacteria and Phages meeting). Finally a second minisession of training courses brings the season to a close in late October.

In the Laboratory's centennial year 1990 a total of 18 training courses were given either on the grounds of the Laboratory or at the Laboratory's off-campus conference facility Banbury Center (see below), with an average of 15 students plus five instructors/assistants participating in each. A total of 12 professional meetings were held in Grace Auditorium with an average attendance of 350 (they ranged in size from 200 to 450 participants). In addition, 14 small meetings and workshops were held at Banbury Center. When the large numbers of guest seminar speakers who lecture to the individual courses are added to this visitors list a total of nearly 4500 scientists visited Cold Spring Harbor to attend courses and meetings during its centennial year.

Ever since the course program was begun in 1945 with the introduction of the Phage Course the philosophy behind the modern training courses at Cold Spring Harbor has remained the same. Each course is chosen to fill the need for instruction in an interdisciplinary field that is either so new or so specialized that it is not feasible for a university to offer such a course. The courses provide such thorough grounding in each specialty that their graduates are able to plunge into work in their chosen new field as soon as the training session is over. Often the success of these graduates is so rapid that it is not unusual to find one or two them lending a hand in teaching the same course only a few years after taking it.

Regardless of the subject, all of the courses at Cold Spring Harbor are small and taught in a workshop-type format by a group of instructors who come together from several different institutions. They are characterized by high teacher-to-student ratios (the one instructor for every three or four students found in the ultramodern cubicles for neurobiology in Jones Laboratory is not atypical) and are greatly enhanced by regularly scheduled seminars given by distinguished visiting lecturers who are well-known researchers in the field. The course laboratories are open from early in the morning until late at night seven days a week, promoting concentrated but informal instruction and study.

Long after the lobster banquet that traditionally signifies the end of a Cold Spring Harbor course the graduates may later converge on the Laboratory to further exchange ideas. Many of the professional meetings held at the Laboratory annually can trace their origins to a specific course, a pattern that was set early on by the Phage Meeting which was first held in 1950, five years after the Phage Course began.

Courses and meetings at the Laboratory generate books—and income

Besides serving as a clearinghouse for scientific knowledge, ever since the first Symposium in 1933 the meetings at Cold Spring Harbor have generated books of immense value to the scientific community, a fact underscored by their success in the marketplace. (Sales of the *Symposia* volumes have always been a reliable source of income for the Laboratory during financially troubled times.) As additional meetings were added to the summer schedule in the early 1970s the possibility for more proceedings volumes arose and the list of books published by Cold Spring Harbor Laboratory grew. The innovative summer courses led to a series of laboratory manuals, the first of which was Jeffrey Miller's *Experiments in Molecular Genetics,* published in 1972, which explained the then brand-new gene selection techniques. Ten years later, in 1982, *Molecular Cloning: A Laboratory Manual* (based on the Molecular Cloning course) first appeared. Now into its second edition (and filling three volumes) this runaway scientific bestseller written by Joseph Sambrook, Edward Fritsch, and Thomas Maniatis has sold over 100,000 copies in its two editions and has been called the "bible" of recombinant DNA research. The eagerly awaited *Antibodies: A Laboratory Manual* by Ed Harlow and David Lane was published in 1988, and Michael Ashburner's instant classic, the 1700-page, two-volume *Drosophila: A Laboratory Handbook and Manual* rolled off the presses the following year. Publishing complex laboratory manuals one after the other in the 1980s as well as producing the large numbers of abstract/program booklets needed for the sizable roster of professional meetings and bringing out the increasingly hefty *Annual Report* of the Laboratory each year would not have been possible had the Publications Department not moved in 1983 from its increasingly cramped quarters in Nichols Building to newly created offices in Urey Cottage, a nearby structure previously used as a residence.

(9) Urey Cottage.

Originally built as a summer cottage in 1934, it was enlarged and remodeled in 1983 to house the editorial and production offices of the Laboratory's Publications Department (now CSHL Press).

(10) Urey Cottage interior, showing books published by the Laboratory.

The 1990 publications list of CSHL Press contained over 140 books, including a number of still-in-print volumes of the world-famous *Symposia on Quantitative Biology* (see top shelf for entire series in their distinctive blood-red bindings).

Renovation and enlargement of
UREY COTTAGE
1983

Built inexpensively in 1934, Urey Cottage grew over the years in fits and starts from a tiny summer cottage to a fully winterized year-round house for Laboratory staff. Having been enlarged in this way, with no overall design in mind, the building gained much in appearance as well as space when it was thoughtfully renovated and enlarged in 1983 for the expanding Publications Department of the Laboratory. The exterior was redesigned along symmetrical Colonial lines gaining a front porch and matching shed-roofed wings at the ends. Ironically now that it was no longer a residence it had become the epitome of the domestic looking neo-"Long Island Colonial" style of architecture which at that time was springing up along Bungtown Road in imitation of the Laboratory architecture of the late 1920s. The editorial and production staff quickly filled the expanded interior of Urey Cottage, which not long after became equipped with sophisticated computer hardware and software so that Cold Spring Harbor books and other publications could be typeset in-house using desktop publishing techniques. *(9,10)*

The Laboratory launches two journals

In addition to the proceedings of many of the professional meetings held at the Laboratory and the laboratory manuals explaining state-of-the-art scientific techniques developed here, Cold Spring Harbor also publishes a well-received ongoing monograph series (begun in 1970) of specially commissioned books on single topics. Over the years the number of books published by the Laboratory has grown steadily and in its 100th birthday year there were more than 140 Cold Spring Harbor Laboratory publications in print. In the late 1980s the book program was augmented by the inauguration of two new scientific journals—*Genes & Development* in 1987 and *Cancer Cells: A Monthly Review* in 1989—with plans for a third, *PCR Methods and Applications*, well in hand by the end of 1990. The idea to publish journals arrived with Stephen Prentis, who came to Cold Spring Harbor in 1986. Following his tragic death in 1987, just days before the first issue of *Genes & Development* appeared, the editorship of this journal passed to staff scientist Michael Mathews; it has been edited since 1990 by Terri Grodzicker, Assistant Director for Academic Affairs. In 1989 the Publications Department became a distinct operating unit called Cold Spring Harbor Laboratory Press (CSHL Press) with its own operating budget. Although the majority of the Laboratory's annual budget is funded by grants from public agencies and private foundations and by funds received from industry and private individuals, the Laboratory's publishing venture is among the most successful of the several ancillary activities at Cold Spring Harbor that together fund upward of 20% of the annual budget. Besides its contribution toward funding science at the Laboratory, CSHL Press now has the added educational mandate to help keep the public, and not just the scientists, informed. Some of the most important Cold Spring Harbor titles these days are generated by the "think tank" meetings, often involving public policy, that are held at the Laboratory's small conference facility, Banbury Center, which was opened in 1977 on the grounds of the former estate of Charles S. Robertson.

An off-campus conference facility, Banbury Center,
takes shape thanks to the Robertson family

Banbury Center was named after Banbury Lane, the road that leads through the 45-acre former Robertson estate. Originating on the west side of West Neck Road (a road dating back to colonial days) Banbury Lane runs due west alongside rolling meadows dotted with stately trees, with homes of earlier and later vintages (under separate ownership) visible in the distance, past an old apple orchard and around a sweeping curve to culminate at the gracious family home built for Charles Sammis Robertson and his wife Marie in the mid-1930s. For hundreds of years this property had been farmed by his mother's forebears, the Sammises, and he had painstakingly reassembled it for posterity.

In the early 1970s Charles Robertson had been a strong financial supporter of the training program in neurobiology inaugurated at Cold Spring Harbor through his establishment of the Robertson Research Fund in 1973 (see Chapter Five; this gift was the Laboratory's first permanent endowment fund). He soon made an additional gift that would have an equally far-ranging effect on the Laboratory's future. With the assurance that it would be preserved through a new use in the interests of science, in early 1976 Charles Robertson deeded to the Laboratory his fine Georgian-style residence and estate in Lloyd Harbor. A year later he saw his wishes fulfilled, for in 1977 the Robertson's Banbury Lane estate became the Laboratory's new off-campus conference facility, Banbury Center, with Victor McElheny, formerly the technology correspondent of the *New York Times,* serving as its first director. At the dedication ceremony Francis Crick, speaking on "Being a Scientist," pointed out the important role that intimate small-scale meetings play in fostering new ideas. The new facility consisted of Robertson's former home, which was named Robertson House in his honor and served as the residence and dining hall for the meeting participants, and his former seven-car garage which was adaptively reused as the Meeting House of the new conference facility.

ROBERTSON HOUSE
(Designed by Mott B. Schmidt;
Built in 1936)
*Donated to the Laboratory
1976*

(11) Aerial view of Robertson House at Banbury Center.

In 1976 Charles Sammis Robertson presented Cold Spring Harbor Laboratory with his home and 45-acre estate in Lloyd Harbor. The buildings on the estate were adaptively reused to suit the purposes of a small conference center, which was formally dedicated the following year as Banbury Center, after its street address, Banbury Lane, off West Neck Road in Lloyd Harbor. (Long Island Sound and Center Island can be seen in the background.)

Sheathed in white-washed brick and situated on a high bluff overlooking Lloyd Harbor, the 26-room Robertson residence was designed by Mott B. Schmidt, a well-known Georgian Revival architect. In addition to designing town and country houses for the Astors, Morgans, Rockefellers, and Vanderbilts (his first large commission was a series of houses at Sutton Place in New York City), Schmidt had been entrusted by New York master planner Robert Moses with the design of the filling stations on the Northern State Parkway, small but elegant stone structures. *(11,12,13,14)*

Credited with the first use of the term "Powder Room" on residential architectural plans, Schmidt himself explained that "the name comes from Versailles, where ladies always resorted to a special chamber to powder their hair." However, he is best known for having come out of retirement in the 1960s to design a suite of reception rooms (later called the Wagner Wing) for Gracie Mansion, the official residence of the mayor of New York.

Spaciousness was a keynote of Schmidt's work; his entrance halls were often 15 feet square. The entrance hall of

(12) Robertson House.

Built in 1936, the house was designed by Mott B. Schmidt, the architect of the 1966 Wagner Wing of Gracie Mansion. The house is articulated as projecting and receding masses, a motif bor-rowed (via English Georgian architects) from sixteenth century Italian architect Andrea Palladio. All of the participants at Banbury Center meetings take their meals here and the majority are also housed here.

(13) Robertson House, detail of rear facade overlooking Cold Spring Harbor.

The Georgian Revival-style door enframement has a broken pediment on top.

(14) View of Cold Spring Harbor from the rear patio of Robertson House.

Robertson House flows into a broad connecting stair hall to one side, the staircase itself being a fine example of the Georgian style with robustly carved balusters. At the far end of the stair hall are a ground floor guest suite and a generously proportioned den with antique paneling, a popular spot for after dinner discussions among meeting participants.

At the opposite end of the entrance hall from the stair hall and den lie the dining room, decorated with hand-painted Chinese wallpaper, and the breakfast room, flower room, pantry, and kitchen. Matching two-story symmetrical wings, their gables, not roof ridges, facing front, are connected by single-story passages to the main block of the house which features a handsome broken-pedimented doorway with dentillated cornice and pediment above. Directly off the entrance hall is the living room, which extends along the back of the main section of the house and affords long water views over the entrance to Cold Spring Harbor from Long Island Sound.

A large home (counting the former house staff quarters there are enough bedrooms to sleep 20), Robertson House is used for dining and lodging purposes. The meeting sessions are held in the estate's former seven-car garage, which is situated 400 yards east of Robertson House and now adaptively reused for this new purpose.

MEETING HOUSE
(Formerly the Garage on the
Robertson Estate;
Designed by Mott B. Schmidt;
Built in 1936)
Transformed by
Moore Grover Harper
1977

Separated from Robertson House by a sweeping lawn and apple orchard, the former Robertson garage is a single-story structure clad in brick and shingles. It was remodeled as the Meeting House of Banbury Center immediately after the transfer of the Robertson property to the Laboratory in 1976. Originally there were three bays for automobiles in the shingled central section of the structure and two each in the brick end sections and their connectors. As transformed by the architects from Moore Grover Harper (antecedent firm to today's Centerbrook), the taller central section of the former garage became the main conference room of the facility, fully equipped for sound and projection. The east wing was renovated as offices for the Center's staff and the west wing became a library and lounge. *(15,16)*

(15) Meeting House at Banbury Center.

The former seven-car garage of the Robertson estate was re-modeled as the Meeting House of Cold Spring Harbor Laboratory's Banbury Center. The tall middle section of the building now houses the main conference room of the facility, and the wings contain offices *(left)* for the Banbury Center staff and a library/lounge *(right)*.

(16) Library/lounge of the Meeting House.

The book collection contains all of the titles in the *Banbury Reports* series as well as standard reference works needed for the neurobiology courses held here during the summer.

After the Banbury Center program gets under way, more beds are needed

Two small meetings on neurobiology, the subject closest to Charles Robertson's heart, were held in the Meeting House at Banbury Center in its inaugural year 1977. In subsequent years meetings of this type were supported by the Marie H. Robertson Fund that had been created in 1975 by gifts from the Banbury Foundation, the Robertson family's philanthropic body. To this day Banbury Center continues to have a strong neuroscience orientation since several of the Laboratory's postgraduate training courses in neurobiology are headquartered here each summer, including an ongoing course on Computational Neuroscience. For this course a full complement of computer setups are stationed around the main conference room of the Meeting House for use by the students in their investigations of the brain through the use of computer modeling. Gradually a program of small specialized meetings held throughout the year began to take shape on the former Robertson estate. Before the Banbury Center program could become truly independent, however, additional on-site housing was needed. Although Robertson House could accommodate 20 visitors, the Meeting House was geared for conferences of 35 to 40 participants. In the early years additional accommodations had to be secured off site.

At first the Laboratory had access to rooms at nearby Fort Hill in Lloyd Harbor, then owned by Anna Matheson Woods, daughter of early Bio Lab president William J. Matheson. Although the Laboratory was subsequently offered this home as a gift, it could not be accepted due to the anticipated high operating expenses. After ownership of Fort Hill was transferred to a nonprofit music organization (The Aspen Foundation) the time had come for the Laboratory to commission plans for building a new residence on the grounds of Banbury Center. Moore Grover Harper was charged with the design of a guest house for 16 visitors.

(17) Sammis Hall at Banbury Center.

Completed in 1981, Sammis Hall provides supplementary housing for visitors at Banbury Center.

SAMMIS HALL
Moore Grover Harper
1981

In keeping with the Georgian style of Robertson House, Charles Moore and his architectural colleagues modeled the new building for the Banbury campus after a country villa designed by Andrea Palladio, one of the foremost architects of the Italian Renaissance. Many of the features associated with the English Georgian style, including total building symmetry and pedimented facades, are in reality Palladian hallmarks of the late sixteenth century. These features were later introduced into England in the early to mid seventeenth century by the architect (and astronomer) Inigo Jones. Two of Jones' most famous buildings are Queens House in Greenwich and St. Paul's Chapel, Covent Garden, London. Jones not only personally examined Palladio's buildings in Italy, but also owned a copy of his great design treatise *Four Books of Architecture,* published in 1570 and illustrated with drawings and floor plans of many of his works.

Charles Moore's Palladian solution to the housing crunch at Banbury Center was a square, two-story stucco (over wood frame) building, predominantly painted a warm terra-cotta color. The tall ground floor window surrounds and the second floor were highlighted in a light, cool, stone gray color. A gravel path lined by two rows of venerable old apple trees leads straight up to the front entrance of the residence hall, defining the main axis of symmetry of the building. *(17,18,19,20)*

(18) Drawing of Sammis Hall dated 1979.

All of the main architectural details of the building as completed two years later were present in this early architectural drawing from the Moore Grover Harper offices in Essex, Connecticut. It highlights the similarity in massing and detail (note the door enframements under the pediments) between Sammis Hall at Banbury Center and the building after which Charles Moore loosely patterned it, the Villa Poiana (see illustration *19*).

(19) Villa Poiana, 1549, Andrea Palladio, architect.

This building served as the model for Sammis Hall, 1981.

(20) Sammis Hall rear facade.

Like the residence that provided its design inspiration—Palladio's 1549 Villa Poiana in Vincenza, Italy—Sammis Hall has a cornice line that rises pediment-like above the arched central entrance. A unique Charles Moore feature on the "Villa Sammis" is the disembodied keystone (no arch) suspended from the apex of the cornice pediment. It is a typically Post-Modern Classical detail. In this style of architecture introduced in the early 1980s (Moore was a key innovator) classically correct features from Ancient or Renaissance architecture are caricatured or otherwise modified, often in highly abstract ways. Below the idiosyncratic Sammis Hall keystone a molded wooden panel, shaped like the missing arch, handily displays the Laboratory's logo inside its curve.

The 16 guest rooms inside are organized as eight two-bedroom suites entered at the four corners of each floor. The center of the building is a monumental space open clear up to the roof. There the base of a cupola intersects the ceiling (painted sky blue). If Palladio himself had drawn the plan of Sammis Hall he would have called this room the Sala, but Charles Moore's unmistakable imprint shines through when the interior of the building is appreciated in three dimensions. A celestially white superstructure of arches soars above the central hall, a glorification of the aedicular motif (a "little house" within a house) that Moore has often used in his domestic designs. *(21, 22,23,24)*

At the dedication of Sammis Hall in the summer of 1981 architect Charles

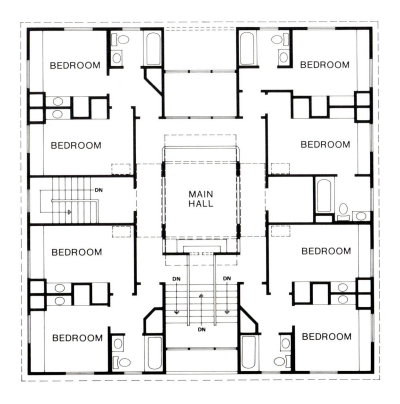

(21) Sammis Hall floor plan, upper level.

The dashed and solid lines surrounding the "Main Hall" denote the piers of the aedicular canopy that floats above this space (see illustration *24*).

(22) "Main Hall" in Sammis Hall.

(23) Sammis Hall interior, looking through entrance loggia.

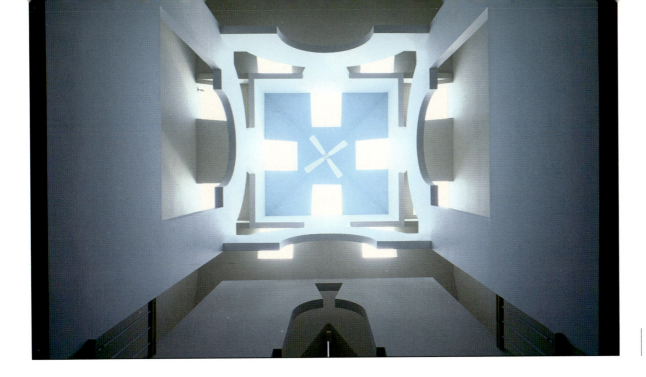

(24) Sammis Hall ceiling detail.

Moore and scientist Gunther Stent gave a pair of complementary speeches on "The Brain of the Architect" and "The Architecture of the Brain." As soon as it was ready the new residence hall, named Sammis Hall in honor of Charles Robertson's forebears, was used to house participants at the late summer meetings held on the Laboratory's main Bungtown campus. Thus by the fall of 1981 the Banbury Center complex was an entirely self-contained conference facility ideal for holding small meetings.

The Banbury Meetings reach a wide audience

By the time the Laboratory's centennial year drew to a close in 1990 Banbury Center had been the scene of over 100 meetings. In contrast to the conferences and Symposia for several hundred scientists held in Grace Auditorium, Banbury Center meetings are small and workshop-like in style. Between 30 and 40 of the world's leading experts in a subject are invited to spend several days together at Banbury Center in an intensive exploration and discussion of their work. At first the Center's program concentrated on issues relating to biological risk assessment, especially the assessment of environmental hazards like carcinogens that act at the genetic level. Early meetings included Assessing Chemical Mutagens (1978), A Safe Cigarette? (1979), Hormones and Breast Cancer (1980), and Genetic Predisposition in Responses to Chemical Exposure (1983), all topics of continuing interest today. In addition, and not surprisingly given the research interests of the Laboratory, Banbury Center is also used for meetings on research in molecular biology and genetics. Especially notable among many significant meetings were those on Recombinant DNA Applications to Human Disease (1982), which foreshadowed the extraordinary advances that have come in human genetics, and Acquired Immunodeficiency Syndrome (AIDS) and Kaposi's Sarcoma (1983), at which the suggestion was first made that AIDS might be caused by a human retrovirus.

Banbury Center also hosts influential meetings on the interactions between biological research and society. Patenting of Life Forms (1981), DNA Technology and Forensic Science (1988), and the Ethos of Scientific Research (1989) all have had an impact that has spread far beyond the confines of the Meeting House. This is due in large measure to the fact that the deliberations and discussions at Banbury Center meetings are disseminated widely through the publishing activities of Cold Spring Harbor Laboratory. The *Banbury Reports* are authoritative accounts of the meetings on environmental issues, whereas papers from those meetings on basic research appear in the *Current Communications in Molecular Biology* (now titled *Current Communications in Cell & Molecular Biology*) series. In addition, through the generosity of the Alfred P. Sloan Foundation, Banbury Center has become a clearinghouse for information on important scientific topics both for congressional aides and for science journalists.

The Long Island Biological Association has been the traditional outreach arm of the Laboratory

Whereas the Laboratory's Banbury Center has had a comparatively brief, albeit far-ranging, history of disseminating topical scientific knowledge to journalists and policy makers as well as scientists, the Laboratory's tradition of reaching out to the Cold Spring Harbor community is a long-standing one dating back to the last century. In the early days of the Biological Laboratory popular talks on such subjects as "Germ Life" and "Oyster Culture in Europe" (summer of 1898) were offered to the neighbors throughout the summer season. Illustrated by lantern slides, the talks were held in the basement lecture hall of Wawepex Building, the former whaling era warehouse. After the formation of the Long Island Biological Association in 1924 these ties with the community became of tantamount financial importance.

For nearly three generations the members of LIBA have been successfully meeting the challenge of raising the funds needed to purchase lands, undertake necessary construction projects, and support research at the Laboratory. Moreover, the Women's Auxiliary of LIBA long ago in 1925 started a Children's Nature Study Program on behalf of the Laboratory, which makes Nature Study the Laboratory's longest-running program of summer instruction still offered. Considering the vast numbers of youngsters who have participated over the years, the Nature Study Program has played an important role in introducing many neighboring parents to the Laboratory.

The Childrens' Nature Study Program is now
headquartered at Uplands Farm

After being headquartered in a series of different structures, including Wawepex Building, the Animal House (now McClintock Laboratory), and Jones Laboratory, the Nature Study Program is now ensconced at nearby Uplands Farm, the former Nichols estate on Lawrence Hill Road that was deeded to the Nature Conservancy in 1974. (Today the Laboratory also maintains a presence there in the form of the Agricultural Field Station of its plant genetics program; see Chapter Six.) The suite of classrooms occupied by the Nature Study Program in the main barn at Uplands Farm was designed especially for its use in the mid-1970s by the architects from Charles Moore Associates. Geared primarily to elementary school children, the field courses offered by the Nature Study Program have such catchy names as "Frogs, Flippers, and Fins," "Nature Bugs," and "Pebble Pubs."

The Laboratory sponsors a DNA Learning Center in
Cold Spring Harbor

Today the chief provider of the Laboratory's educational outreach in terms of high school students and their teachers is the DNA Learning Center founded in Cold Spring Harbor in 1988. It was the brainchild of David Micklos, a professional educator who first came to the Laboratory as its public information officer late in 1982. In 1985 he founded a DNA Literacy Program under the Laboratory's sponsorship with the purpose of bringing local high school biology teachers up to speed on the kind of science that was being done at Cold Spring Harbor so that they could share it with their students. The DNA Literacy Program went national a year later after a grant from Citibank made possible the purchase and equipping of the first of two Vector Vans (the second was acquired in 1987). These mobile DNA laboratories enabled program staff to travel to schools across the country and teach recombinant DNA techniques to high school teachers in intensive hands-on summer workshops. Although a hit on the road, the DNA Literacy Program still lacked a home base in Cold Spring Harbor.

Then late in 1987 a former school building in Cold Spring Harbor became available for rent (with an option to buy). In the fall of the following year the DNA Learning Center of Cold Spring Harbor Laboratory was officially opened in this Georgian brick structure at the east end of Main Street that had originally been built as the Union Free School of Cold Spring Harbor. Thus the story of the DNA Learning Center, like those of Banbury Center and many of the buildings on the main Laboratory campus, is a case study of the successful adaptive reuse and architectural preservation of an existing structure.

DNA LEARNING CENTER
(Formerly Union Free School of
Cold Spring Harbor
Designed by Peabody, Wilson & Brown;
Built in 1925)
*Renovated
1988*

The new home that the DNA Literacy Program occupied in 1988 was a 1925 schoolhouse originally designed by the well-known New York architectural firm Peabody, Wilson & Brown, the same architects who completed the plans for the 1914 Animal House (now McClintock Laboratory). Able interpreters of whichever style a given institutional situation called for, the design they produced for the school building at the foot of Goose Hill Road in Cold Spring Harbor was Georgian Revival in flavor, in keeping with its public character. Built in the shape of a wide, but not deep, squared off "U," the building featured matching end pavilions accented by big bull's-eye windows in their gables. *(25,26, 27,28)*

Like many other institutional buildings designed by Peabody, Wilson & Brown, the Union Free School featured a square cupola resting authoritatively on the roof ridge. The firm also designed the Cold Spring Harbor Library, built at the west end of the village in 1913, which sported an octagonal cupola. After several new schools were built in the area the Union Free School became the school district's administrative offices. Then late in 1987 when both Cold Spring Harbor's library and its school administrative offices moved into a redundant 1950s school structure on Goose Hill

Road at the east end of the village, the former schoolhouse building was made available to the DNA Literacy Program. (The library edifice was later purchased by the Society for the Preservation of Long Island Antiquities, which opened it to the public in late fall of 1990 as a gallery for displaying changing exhibits of Long Island decorative and fine arts.) Grants to the DNA Literacy Program from the Josiah Macy, Jr., Foundation and the National Science Foundation, as well as major gifts from over a dozen other sponsors, both private and public, enabled work to begin on adapting the school building to its new use.

Even before the building was formally reopened to the public in 1988 as the DNA Learning Center it was already receiving its first visitors on a regular daily basis as classes of high school biology students performed state-of-the-art experiments in a specially designed and constructed *Bio2000* Laboratory on the main floor. Today a variety of different laboratory exercises are offered, all specially developed at the DNA Learning Center and geared to student audiences. Among the favorites are those involving DNA restriction analysis, bacterial transformation, and the polymerase chain reaction. The basement area of the building contains offices for the DNA Learning Center's staff as well as a testing

(25) DNA Learning Center in Cold Spring Harbor.

In the fall of 1988 Cold Spring Harbor Laboratory opened its DNA Learning Center on the premises of the 1925 Union Free School of Cold Spring Harbor. High school biology students and their teachers come here today to learn the latest DNA experimental techniques. The Learning Center is also open to the public on weekends and weekdays (except Mondays) for viewing several informative exhibits and visiting the well-stocked bookstore.

(26) Union Free School as built in 1925.

It was designed by Peabody, Wilson & Brown, the architects of the former 1914 Animal House, now McClintock Laboratory, on the main campus of Cold Spring Harbor Laboratory. Immediately before becoming the DNA Learning Center it had housed the administrative offices of the local school district. (Photograph from *Architectural Record*, July 1930, p. 44.)

(27) Union Free School floor plan.

This plan from *Architectural Record* (July 1930, p. 44) shows the original uses of the interior space. (Compare with illustration *28*.)

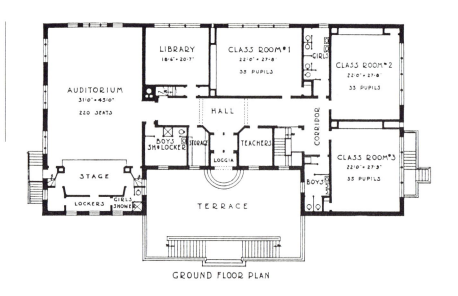

GROUND FLOOR PLAN

laboratory where experimental procedures are worked out ahead of time and a kitchen where materials are prepared for the daily laboratory sessions. (*29*)

After the *Bio2000* Laboratory had been installed in a former classroom, the former school auditorium space was remodeled for use as a large exhibit hall. When the DNA Learning Center was formally opened in September of 1988 the public was treated to a showing of "The Search for Life: Genetic Technology in the 20th Century," an exhibition that originated in Washington, D.C., at the National Museum of American

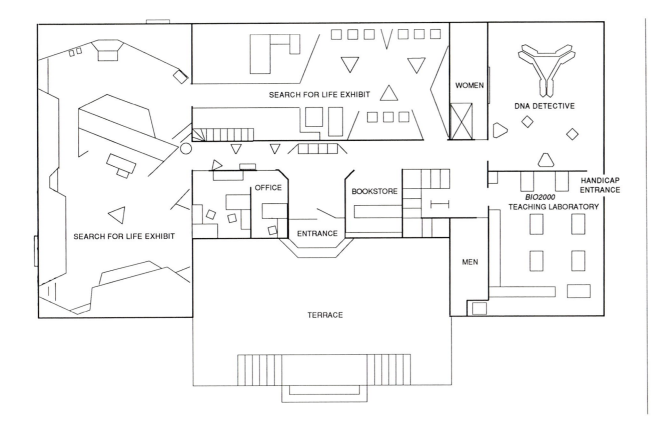

Labels in floor plan: SEARCH FOR LIFE EXHIBIT · WOMEN · DNA DETECTIVE · SEARCH FOR LIFE EXHIBIT · OFFICE · BOOKSTORE · HANDICAP ENTRANCE · BIO2000 TEACHING LABORATORY · ENTRANCE · MEN · TERRACE

(28) DNA Learning Center floor plan.

The "Auditorium," "Library," and "Class Room #1" of the Union Free School (see illustration *27*) were remodeled as one large exhibition space. Here the Smithsonian Institution's "Search for Life" exhibit (see illustration *30*) was staged to coincide with the opening of the DNA Learning Center in 1988. Earlier that year "Class Room #3" had been converted to the "*Bio2000* Teaching Laboratory" (see illustration *29*) for high school biology students. In 1989 an interactive exhibit called "DNA Detective" (see illustration *31*) was mounted in the former "Class Room #2."

History of the Smithsonian Institution. Cold Spring Harbor Laboratory was closely involved in developing this multimedia presentation on the history of DNA discoveries, in which the key contributions of several Cold Spring Harbor scientists are highlighted, and the Laboratory supplied numerous historical artifacts and photographs as well as video footage for the exhibition. On the afternoon of the formal opening of the DNA Learning Center guests were invited to tour the building, and back at Grace Auditorium there was a program of several short speeches plus a keynote address, "Reading DNA," by Robert Pollack, Dean of Columbia College and a former staff scientist at the Laboratory. By serving as the world's first museum of DNA the

Learning Center plays an important role in interpreting to the layman the kind of science that is done across the harbor at the Laboratory, where public access is more limited. *(30)*

Immediately adjacent to the *Bio2000* Laboratory another former classroom was transformed into a smaller exhibition space where an interactive "DNA Detective/DNA Diagnosis" exhibit mounted in 1989 demonstrated the usefulness of the DNA fingerprinting technique in resolving animal breeding and paternity cases and in investigating crimes. This technique is based on identifying the specific regions where the DNA that composes a chromosome varies sharply from individual to individual—the ultimate expression of in-

(29) *Bio2000* Laboratory at the DNA Learning Center.

Laboratory exercises offered here to prebooked classes of biology students feature DNA restriction analysis, bacterial transformation, and the polymerase chain reaction.

(30) Smithsonian exhibition, "The Search for Life: Genetic Technology in the 20th Century," mounted at the DNA Learning Center.

Cold Spring Harbor Laboratory was closely involved with the Smithsonian Institution in developing this multimedia presentation on the history of DNA discoveries.

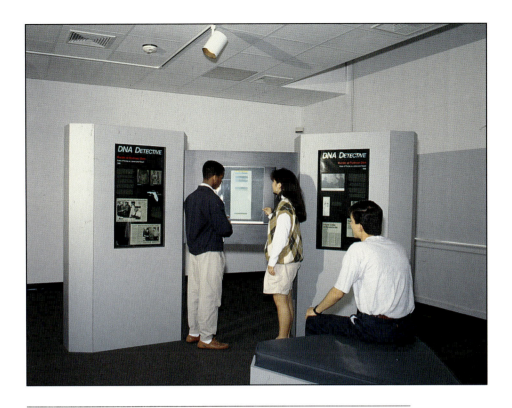

(31) "DNA Detective/DNA Diagnosis" exhibit at the DNA Learning Center.

This interactive exhibit focuses on the usefulness of the DNA fingerprinting technique in forensic medicine and genetics research.

dividual identity. Unlike traditional fingerprinting, the patterns revealed by DNA fingerprinting are a unique blend of those of the two parents of the individual. *(31)*

However, as its name implies, the DNA Learning Center is much more than a museum or exhibition hall or even an adjunct to school-based biology courses. Regularly scheduled workshops are held at the Center for biology teachers who wish to learn how to interpret the new science of DNA through innovative laboratory exercises and also how to equip the science laboratories at their home schools for executing the same kinds of experiments they see demonstrated in Cold Spring Harbor.

Reflecting the rapid pace of DNA science today, the DNA Learning Center's constituency is expanding to include students not only at the high school level, but also at the middle and elementary school levels. Its influence has also percolated upward to the highest eschelons of business, commerce, and politics through the custom-tailored visits to the DNA Learning Center arranged for these groups.

*The Laboratory helps train the next generation's
research scientists*

In addition to educating schoolchildren, their teachers, and the general public at the
DNA Learning Center, the Laboratory has long provided special educational opportunities at
its main campus on Bungtown Road aimed at fostering future scientific talent and training
the next generation of researchers. At a distinctly more advanced level than the ecology-
oriented Nature Study Program, the Laboratory's Undergraduate Research Program designed
for college undergraduates who are contemplating careers in biomedical fields was inaugurated in 1959 with initial financial support coming from the Rockefeller Foundation. Each
summer 18 to 20 students, nicknamed "URPS," are selected from colleges and universities
in the United States and Europe to spend a ten-week period at the Laboratory during the
height of the summer season doing research in conjunction with scientific staff members.

Modeled on this long-running URP program is a new program for high school students inaugurated in the Laboratory's centennial year. Called Partners for the Future it runs
for 20 weeks during the school year and brings selected high school seniors into the laboratories. Sponsored by Long Island businesses in partnership with Cold Spring Harbor Laboratory, this newest scheme, like the URP program, is aimed at encouraging students to consider professional careers in science, but at an even younger age.

*Support for science at Cold Spring Harbor is
increasingly coming from the private sector*

Although the Laboratory has received gifts from corporations, individuals, and private foundations that have helped create Banbury Center and the DNA Learning Center as
well as upgrade the facilities along Bungtown Road, since the late 1960s it has depended on
grants from the federal government to run the actual scientific programs. Toward the end of
the period of the most rapid expansion of the Laboratory, 1970-1990, however, the relative
amounts of public versus private funds underwent a significant change. Out of the Laboratory's total income in 1970 of approximately $1.41 million, grants from the federal govern-
ment amounted to approximately $830,000 (of which the Cancer Center grant was by far the
largest) or nearly 60% of all of the available funds. The total number of employees that year
was 80, of which 35 were professionally trained scientists. In 1990 there were 450 Laboratory
employees, of which 158 were scientists. The budget that year was $28 million, of which
slightly less than 40% represented income from federal granting agencies. Approximately 20%
of the 1990 budget came from auxiliary activities, particularly the publication of books generated by the training courses and professional meetings. Corporate contributions, endow-

ments, and interest plus miscellaneous other income each accounted for roughly 5% of the budget. Next to federal grants, however, the second largest source of income to the Laboratory, accounting for more than 25% of the budget, was grants from private foundations and private contributions.

*The Laboratory launches its Second Century
Campaign...and briefly pauses to celebrate its 100th
birthday*

Focusing on the increasingly important private sector, the Laboratory began planning in 1988 for its Second Century Campaign, with a goal of raising $44 million. The campaign had three specific financial objectives: to build up the Laboratory's endowment, to create a fund for the improvement and maintenance of infrastructure facilities, and to raise the funds needed to build a Neuroscience Center at Cold Spring Harbor. Laboratory treasurer David L. Luke III spearheaded the Second Century Campaign, and to assist the fund-raisers a Development Office was opened in the basement of Wawepex Building in the spring of 1988.

Under the direction of newly appointed Public Affairs head (and long-time Library Services director) Susan Cooper, planning also began at this time for celebrating the Laboratory's 100th birthday in July of 1990. The inaugural event of the Cold Spring Harbor Laboratory Centennial was the dedication of the DNA Learning Center in the fall of 1988. Other centennial events leading up to the birthday celebration included the laying of the cornerstones for the buildings of the Neuroscience Center in the spring of 1989 (see below) and the dedication of the New Cabins in the summer of 1989. In December of 1989 Jonas Salk addressed Laboratory members and friends on the subject of "HIV and a Possible AIDS Vaccine," and in January of 1990 an art exhibit entitled "Davenport's Daughter" opened in Bush Lecture Hall; on display were paintings, sculptures, and woodcuts by Jane de Tomasi, daughter of Laboratory director Charles Davenport and widow of Director Reginald Harris.

The annual Symposium in the Laboratory's centennial year was on The Brain, a most appropriate choice as the Neuroscience Center rapidly took shape on the hillside above Grace Auditorium. Coinciding with the Symposium every year in June a Dorcas Cummings Lecture is held for the membership of the Long Island Biological Association, and in 1990 Francis Crick, distinguished alumnus of many Cold Spring Harbor Symposia, addressed a standing-room-only audience of LIBA members and friends plus fellow Symposium participants on the subject of "How Do We See Things."

A short six weeks later Centennial Day dawned on July 14, 1990, heralded by the arrival at Cold Spring Harbor of several "tall ships" which later in the day were made avail-

(32) Re-creation of the first biology class at Cold Spring Harbor on the occasion of the Laboratory's 100th birthday on July 14, 1990.

Volunteers in authentic late nineteenth century attire steamed toward the dock at Jones Laboratory in a replay of an 1890 collecting trip by a group of Cold Spring Harbor's first summer biology students that was captured for posterity in an old photograph (see Chapter 1, illustration 6).

able for boarding. Picnicking on a grand scale was the keynote of the afternoon, solemnized by the reenactment (repeated once) of a waterborne collecting trip by the Bio Lab's first biology class in 1890. Gliding across the harbor on a beeline for Jones Laboratory came a replica of the Laboratory's first boat, the "Rotifer," bearing studious-looking mates, class of 1890, dressed in period costumes to match those of the first students (replicated from the early photograph shown in Chapter One, illustration 6). This moving re-creation of an historic moment for Cold Spring Harbor capped off a day filled with both feasting and music, including sea chantey singers from Mystic, Connecticut, and the brass band from Long Island's own Old Bethpage Village Restoration. (32)

After the formal presentation of numerous proclamations commemorating Centennial Day at Cold Spring Harbor Laboratory—it was recognized by the town of Huntington, the counties of Suffolk and Nassau, the state legislature of New York, and even the president of the United States, George Bush—the evening's activities went into full swing. A buffet dinner was served to 400 local business leaders and guests on the grounds of Airslie, the current Laboratory director's residence, while more concertizing got under way back at the Laboratory's historic hub, the sweeping lawn behind the first director's residence, Davenport House. Soon the evening sky over the harbor was illuminated by the first brilliant bursts of a matchless display of pyrotechnics by Grucci, "the first family of fireworks," without whom no centennial would be complete. The misty low clouds over Cold Spring Harbor on that evening magnified the Grucci's spectacular works, which were enjoyed not only by the Laboratory's staff and friends hunkered down on the Bungtown waterfront, but also by scores of neighbors across the harbor and by a myriad of boaters anchored offshore.

As the evening came to a close and the dawn of the second century of Cold Spring Harbor Laboratory lay just over the horizon the crowning achievement of the centennial celebration still lay months ahead, the dedication of the Laboratory's new Neuroscience Center slated for the spring of 1991. This impressive facility for research and teaching in the neurosciences was the single most important focus of the Second Century Campaign and fund-raising for this continued to be successfully accomplished behind the scenes by the dedicated trustees and other generous volunteers even while the Laboratory's birthday celebration stole center stage.

The neurobiology tradition at Cold Spring Harbor
stretches back in time

By the late 1970s and early 1980s the need for a year-round neurobiology facility had become increasingly apparent as the world of traditional neurophysiology now increasingly interfaced at Cold Spring Harbor with the world of recombinant DNA technology. The summer neurobiology program at Cold Spring Harbor had begun in 1971 as an effort to encourage scientists trained in other fields, especially molecular biologists at loose ends after the elucidation of the genetic code in 1966, to explore the world of neurobiology, which up until that time was known as neurophysiology and had a strong component of traditional biophysics. The latter field was not new to Cold Spring Harbor; much of the research at the Bio Lab in the 1930s had been focused on the electrical properties of cells, specifically on the functioning of the outer cell membranes and their role in the conduction of nerve impulses. Even as recently as 1965, just one year before the Symposium on The Genetic Code, Sensory Receptors had been the topic at this annual meeting. Using a 1970 start-up grant of $250,000 from the Sloan Foundation, which had just inaugurated a grant policy of awarding large gifts for the development of academic centers in neurobiology, the Laboratory made plans to offer two initial summer neurobiology courses and to renovate the laboratory space necessary to do this. As a result the 1914 Animal House was rejuvenated, gaining a new teaching laboratory and a library/seminar room on the second floor, and renamed McClintock Laboratory (see Chapter Five).

Both a lecture course, Basic Principles of Neurobiology, and a laboratory-based course, Experimental Techniques in Neurobiology, were offered that first summer, 1971. The former was taught by a group from Harvard Medical School and the latter by the husband and wife team from Paris Philippe Ascher and JacSue Kehoe. In their laboratory course Ascher and Kehoe introduced the use of electrophysiological methods by working with *Aplysia californica,* a large mollusc with nerve ganglia big enough to be pierced with microelectrodes and thus to have their electrical potentials recorded. A third course was added in the summer of 1972 on the Behavioral Genetics of a Nematode, specifically *Caenorhabditis elegans,* an experimental organism that had already won the allegiance of molecular geneticist Sydney Brenner. Several of his colleagues, all former members of the Phage Group (see Chapter Four), had already entered the neurobiology field. These included Seymour Benzer, who had adopted the fruit fly *Drosophila melanogaster* as a model system, and Gunther Stent, who was experimenting with the European medicinal leech *Hirudo medicinalis*. In 1973 and 1974 new courses added to the Cold Spring Harbor neurobiology roster focused on the fruit fly and the leech, the former taught by William Pak from Purdue University and the latter by John Nicholls, who had moved from Harvard Medical School to Stanford University.

Neurobiology briefly becomes a year-round as well as summertime endeavor

The single strongest positive impact on the future of neurobiology at Cold Spring Harbor was the munificence of Charles Robertson and his family, which first began to be felt at the Laboratory in the mid-1970s. Monies from the Robertson Research Fund were used in 1975 for the renovation of Jones Laboratory as a second teaching laboratory for the neurosciences (see Chapter Five). Later the opening of Banbury Center on the grounds of the Robertson estate in the summer of 1977 meant that there was now room for additional lecture courses. An Ad Hoc Neurobiology Committee was assembled at the end of that summer to explore how the Laboratory could best utilize its now considerable physical resources for teaching neurobiology and to look into the possibility of reintroducing year-round research in this field.

In 1978 Ad Hoc Committee member Eric Kandel from Columbia University College of Physicians & Surgeons taught a course at Banbury Center on the Neurobiology of Behavior for the first time; the total number of neurobiology courses offered by the Laboratory that summer was six, together with five small advanced workshops. The formal courses were chiefly funded by training grants from the National Institutes of Health, an exception being the new course on Techniques for Studying the Vertebrate Central Nervous System that focused on the cat visual cortex and was introduced by Harvard Medical School-trained Carla Shatz and funded by the National Science Foundation. In addition, a grant from the Esther A. and Joseph Klingenstein Fund of New York made possible the introduction of a new course on Basic Neuroanatomical Methods. The various workshops were supported by monies from the Marie H. Robertson Fund and grants from the Rita Allen Foundation and the Sloan Foundation.

Year-round research, which had been given a firm go-ahead the previous summer by the members of the Ad Hoc Neurobiology Committee, was re-inaugurated at Cold Spring Harbor in the fall of 1978 when Birgit Zipser, who had been a student in the 1975 Neurobiology of the Leech course, arrived from Downstate Medical School to continue her research on the nerve cells of the leech. She was soon joined by Ronald McKay, who previously had been making monoclonal antibodies against oncogenic proteins in James Laboratory. Focusing on the leech, Zipser and McKay were able to produce 40 different antibodies, each recognizing a different set of leech neurons. To help exploit their results they invited electron microscopist Susan Hockfield, then at the University of California at San Francisco, to join them at Cold Spring Harbor.

This collaboration based on using the latest techniques of molecular biology to try to find the molecules underlying the development of the nervous system was so very successful that ironically it spelled the end of the Laboratory's year-round neurobiology effort practically in its infancy. The facilities available at Cold Spring Harbor—or rather the facilities NOT available here during the summer months when the teaching program was in full swing—simply could not compete with those soon offered to these three innovative experimentalists by major universities. Hockfield's departure for Yale Medical School in 1985 signaled the end of year-round neurobiology research for the time being. This cloud had a silver lining, however, inasmuch as Hockfield agreed to continue to organize the summer neurobiology program at Cold Spring Harbor, and her tenacious efforts in grant writing helped procure the equipment needed to bring recombinant DNA techniques into the various courses. Although the summer program was secure, it was eminently clear that before another attempt could be made to launch a year-round neurobiology research program a laboratory specifically designed for this field had to be built.

Planning begins for a year-round neurobiology laboratory

Plans were soon being drawn up by Centerbrook, the architects of Grace Auditorium, for an Adirondack-style, two-story laboratory building of 16,000 square feet to be erected together with a heated visitor residence hall of a similar style on the large level site west of James Laboratory. The site was then occupied by Page Motel (later demolished) and edged by the old Cabins (two of which were subsequently torn down), unheatable accommodations that dated back to the 1950s and were becoming increasingly unsuitable now that the visitor season at Cold Spring Harbor included spring and fall. It was estimated that the laboratory building would cost five million dollars. Before announcing a fund drive for such a large sum, however, several major gifts would have to be in hand. The Lucille P. Markey Charitable Trust, which was approached first, could not then help with capital construction costs for a new laboratory because of a policy to that effect. A subsequent request to the Markey Charitable Trust for the financial help needed to create an X-ray diffraction group at Cold Spring Harbor was granted in 1986, permitting a sizable investment to be made in X-ray crystallography equipment and the professional staff required to run it.

Two major grants of bricks and mortar funds for building the neurobiology laboratory were received the following year in rapid succession, one million dollars each coming from the Klingenstein Fund (that had earlier helped finance the summer neurobiology teaching program) and from the James S. McDonnell Foundation of St. Louis. The Howard Hughes Medical Institute, about to begin making capital gifts once negotiations with the Internal Revenue Service were completed, was also approached at this time for a multimillion dollar grant. Eventually this Institute awarded five million dollars toward constructing and equipping teaching laboratories in the new building as well as one million dollars each for staff support and the summer course program. However, funds were still needed for the research laboratory component of the building. In addition, the X-ray crystallography group was already outgrowing their semi-improvised quarters in Hershey Building and Demerec Laboratory, and the new neuroscience building was the logical place for them to move. To provide this additional space, however, the new building would have to be larger.

Another important consideration before the designs for the neurobiology facility (and the necessary adjacent visitor residence hall) could be finalized was that as Cold Spring Harbor Laboratory approached its 100th birthday in 1990 perhaps the time had come to officially retire the notion that it was still a "summer camp for biologists." All of the evidence now pointed to the fact that it was here to stay; the completion of impressively solid Grace Auditorium in 1986 had provided the most recent and obvious proof. The design of the neurobiology laboratory was thus thoroughly reconsidered and amplified, the thought being that an "angel" for the kind of neuroscience facility that the Laboratory really needed just might appear, possibly in the shape of Dr. Arnold Beckman, a personal friend of one of the Laboratory's trustees and a noted physical chemist and entrepreneur (he was the founder of Beckman Instruments). An attractive model for a "Neuroscience Center" was made in time for a visit to Cold Spring Harbor by Dr. Beckman, who had already expressed interest in the project contingent upon the Laboratory's receiving the Hughes money.

The Centerbrook architects (William Grover and James Childress, collaborating with Charles Moore) had revamped the design for the new buildings along more ambitious lines, and the model that took shape from their new set of schematic drawings dramatically set out the new components of a multistructure Neuroscience Center. No longer to be built with logs, the buildings would be clad in brick and distinctly larger and taller. To diminish their apparent size they were articulated as smaller discrete masses gathered around a courtyard that would be graced by a freestanding bell tower, or campanile. As soon as Dr. Beckman's intention to aid the neurobiology project with a gift of four million dollars was made known early in 1988, working drawings were immediately begun on what was thereafter to be called the Arnold and Mabel Beckman Laboratory. These were followed by the drawings for what was soon to become Dolan Hall, together with the specifications for the campanile that would bear the name Hazen Tower. The building contractors chosen were A.D. Herman Construction Company, who previously had built Grace Auditorium.

Ground was broken for the Neuroscience Center in the fall of 1988, by which time two more major grants had been received. The W.M. Keck Foundation of Los Angeles made a gift of two million dollars to fund construction of the W.M. Keck Structural Biology Laboratory, a facility for X-ray crystallography to be located on the basement level of Beckman Laboratory. Likewise the Dolan Family Foundation of Oyster Bay donated two million dollars to be used toward the construction of Dolan Hall, an overnight residence for 60 visitors. Before work could begin on the buildings, however, the first phase of the project had to be completed, the erection of a 150-car parking garage (it would be used initially by workmen on the site) to be nestled into the hill at the extreme western edge of the site; built on two levels, it would have covered parking in the basement and open-air parking on top. Construction of Beckman Laboratory and Dolan Hall began in May of 1989, an event heralded by the ceremonial laying of the cornerstones followed by a kickoff dinner for the Second Century Campaign. All the monies needed to complete the project were on hand by June of 1990—a total of $21.5 million—the last gift needed to push the fund-raising goal over the top being a generous donation from the Ira W. DeCamp Foundation.

Another important gift that helped greatly was a grant of four million dollars from the Markey Charitable Trust. (A new grant policy now made possible this award of two million dollars to complete the outfitting of the individual laboratories of the Neuroscience Center as well as two million dollars for scientific equipment and staff development.) In 1989–1990 major foundation gifts came from the Pew Charitable Trusts, the Kresge Foundation, the Fannie E. Rippel Foundation, the Bodman Foundation, the Achelis Foundation, the Booth Ferris Foundation, the Nichols Foundation, and the Elaine E. and Frank T. Powers, Jr., Foundation. Donations were also received from private individuals including large gifts from friends, neighbors, and trustees of Cold Spring Harbor Laboratory such as Amyas Ames and Pauline Ames Plimpton, Robert B. and Mary R. Gardner, George W. and Lucy P. Cutting, Taggart and Katharine Whipple, Dr. Laurie J. Landeau and Family, and Wendy VanderPoel Hatch. A major gift was received from Lita Annenberg Hazen of New York in whose honor Hazen Tower was named.

Completion of the Neuroscience Center was slated for spring of 1991, and to ensure that this happened on time an early May dedication date was set long in advance. One week before the formal dedication Dr. Arnold Beckman paid a special visit to Cold Spring Harbor for a ribbon-cutting ceremony in the presence of the Laboratory's trustees, thus officially opening the building on the occasion of the spring board meeting which was held in the beautiful large seminar room on the ground floor of Beckman Laboratory, overlooking Bungtown Road and the head of the harbor. On the afternoon of the formal dedication ceremony, May 3, 1991, the roster of speakers included the Reverend T. Carleton Lee, Rector of St. John's Church (located just across the road from Cold Spring Harbor Laboratory and founded by the same family who had founded the Laboratory, the Joneses), who offered an invocation; Laboratory Board of Trustees chairman Dr. Bayard D. Clarkson, who in his introductory speech outlined the history of how the Neuroscience Center came to be; Laboratory director James D. Watson, who made opening remarks about the facility from the user's point of view; and Mrs. Marianne Dolan Weber, who commented on the importance

(33) Neuroscience Center aerial view.

In this view looking northeast across Cold Spring Harbor, Hazen Tower (bell tower) and Beckman Laboratory (neurobiology research and teaching laboratory) are to the left and Dolan Hall (residence hall) is to the right. The concrete ring directly below Hazen Tower is the light well for the basement level parking. (The New Cabins are visible in the lower left-hand corner, and the Blackford Hall addition can be seen under construction in the lower right-hand corner.) (See also illustration *34*.)

of Cold Spring Harbor Laboratory from a donor's point of view. These remarks were followed by a series of short scientific talks—"The First Neurobiology Course at Cold Spring Harbor Laboratory" by Dr. Stephen Heinemann of the Salk Institute, a student in the first neurobiology course taught here in 1971; "Neurobiology at Cold Spring Harbor Laboratory" by Dr. Susan Hockfield of Yale University, director of the summer neurobiology training program; and "The Future of Neurobiology" by Dr. W. Maxwell Cowan, Vice President for Research and Chief Science Officer of the Howard Hughes Medical Institute, a member of the Neurobiology Advisory Committee and a trustee of the Laboratory.

The afternoon's program drew to a close with a lively dedicatory address by keynote speaker Representative Robert J. Mrazek of the Third Congressional District, who after reminiscing about his summer vacations spent behind a lawn mower on the grounds of the Laboratory went on to pledge his support for greater funding for biological research. A member of the House Appropriations Committee, Representative Mrazek remarked in reference to the Laboratory's choice of neurobiology as its next major field of endeavor, "I can think of no better direction or priority for Cold Spring Harbor Laboratory as it enters its second hundred years.... With the type of boldness and daring that launched our space program 30 years ago, the Neuroscience Center will be tackling what its founders are legitimately calling the ultimate challenge for biologists." Following the ceremony a reception was held in the courtyard of the Center, and although the afternoon of that long-anticipated day was a seasonably chilly one, the handsome new buildings and the highly informative speeches by Cold Spring Harbor's neurobiological alumni and friends were warmly received by its neighbors and special guests. *(33,34)*

(34) Neuroscience Center site plan.

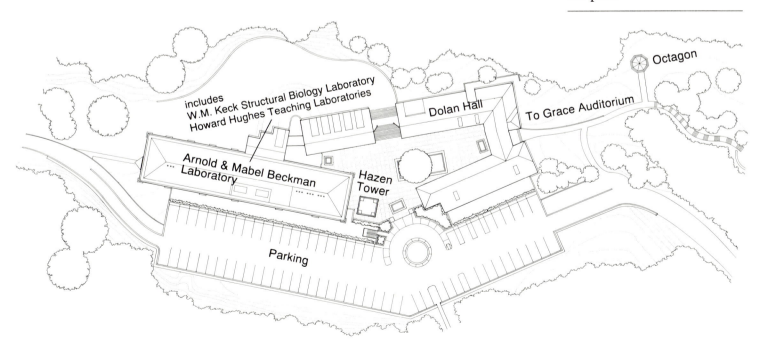

(35) Neuroscience Center from parking area.

Hazen Tower is in the foreground, Beckman Laboratory to the left, and Dolan Hall to the right.

BECKMAN LABORATORY
Centerbrook
1991

Beckman Laboratory is the dominant element in the design of the Neuroscience Center and occupies the northern end of the site. It is composed of a large rectangular laboratory wing, sheathed in dark burnt-red brick, running north and south, and a smaller office/seminar wing, clad in tan brick, angled away from the laboratory in a southeasterly direction. The total area of the two upper floors of the building and the lower level is nearly 38,000 square feet. (There is also a "penthouse" floor on the top of the building that is taken up completely with mechanical equipment.) At the vertex of the angle that the two wings form coming together at the north end of the building is the granite entrance portal that serves both the laboratory and office wings. *(35,36)*

When viewed from across the harbor the main laboratory wing appears to be smaller than it really is due to its dark brick exterior which causes it to recede into the hilly wooded backdrop. It is also partially hidden by the office wing positioned in front of it from this east-facing perspective. The office/seminar wing, itself designed with extra large windows that make it appear smaller when viewed from a great distance, could be mistaken for a grand, waterview-endowed Long Island mansion designed in classical turn-of-the-century style. (This architectural sleight of hand is also true of Dolan Hall, the residential structure that occupies 28,000 square feet on the south end of the Neuroscience Center site

and is sheathed in the same light tan bricks as the seminar wing of Beckman Laboratory; see below.) *(37)*

The bricks used on the facade of Beckman Laboratory (and the two other concrete buildings comprising the Neuroscience Center) were laid up in imitation of Flemish bond as opposed to the more commonly used running bond method of bricklaying that denotes simply running bricks end to end so that only their long sides are visible. Flemish bond means alternating their long sides (stretchers) with their short ends (headers) to produce the more complicated, pleasing, and solid look associated with buildings of the old world. This same effect can be achieved at lower cost by using a special twelve-inch brick that has a groove dividing it into an eight-inch "stretcher" section and a four-inch "header" section, the latter scored to look darker as the bricks chosen for headers often are. When bricks fabricated in this way are laid up in running bond, as they were in the Neuroscience Center structures, they look like the more painstaking Flemish bond. Besides this specially grooved type of pseudo-Flemish bond brick, 15 other brick shapes were used in executing the details called for in the design of the buildings of the Neuroscience Center. The bricklaying for the project was a formidable task requiring the presence of 40 masons on the site at the appointed time.

At the same time much work was occurring on the interior of Beckman Lab-

(36) Beckman Laboratory from the south.

The two main floors of the building to the left each have a Howard Hughes Teaching Laboratory as well as three research laboratories. Behind the green granite entrance portal at the center are the lobby and offices for the scientific staff. To the right is the seminar wing. (Dolan Hall is in the left and right foreground.)

(37) Beckman Laboratory from the east.

This view from Bungtown Road shows the copper-inlaid exterior "buttresses," a neo-Collegiate Gothic touch. Dolan Hall is to the far left (with the shed-roofed dormer) and immediately adjacent to it on the right is the similarly clad (in light tan bricks) seminar wing of Beckman Laboratory; the main part of the Laboratory is to the right (sheathed in dark reddish-brown bricks).

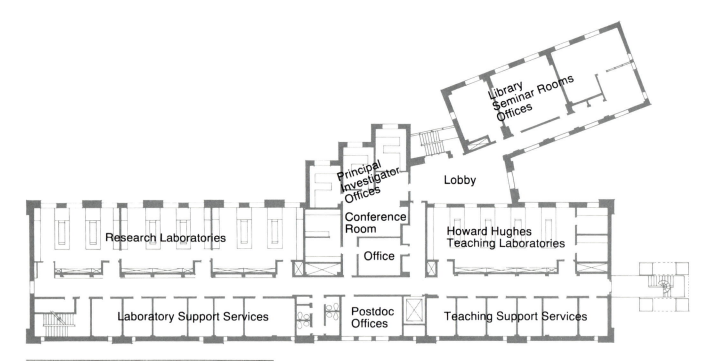

(38) Beckman Laboratory first and second floor schematic plan.

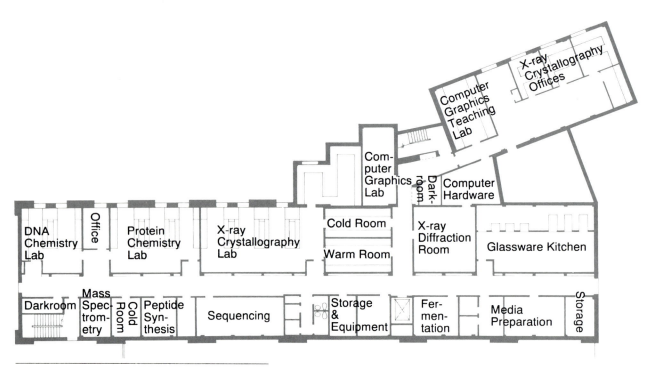

(39) Beckman Laboratory lower level plan.

oratory, the final phase being the construction and furnishing of the individual laboratories and their ancillary spaces. To maximize the synergy between the research and the teaching activities that would take place in the new building, both the first and second floors of Beckman Laboratory were designed to contain a mix of three large research laboratories for teams of six to eight scientists, together with a Howard Hughes Teaching Laboratory for course work and all the necessary support facilities such as equipment rooms, cold rooms, dark rooms, cell culture rooms, and media preparation rooms. The first major equipment order for Beckman Laboratory was appropriately one for $436,000 worth of centrifuges from Beckman Instruments, the highly successful company founded on the strength of one important invention, the pH meter, which Dr. Beckman himself initially called an "acidometer."

In the area at the juncture of the main section of Beckman Laboratory with the smaller angled wing are offices and nearby coffee areas for scientific staff. The seminar rooms and other facilities are in the wing itself. On the ground floor is the spacious Plimpton Seminar Room, which opens directly onto the courtyard and the adjoining Whipple Lounge. On the floor above are Gardner Library, Pulling Seminar Room, and the Director's Office. The public spaces in the seminar wing are all graced by lighting fixtures of a unique design from the Centerbrook architects' own drawing boards. Below the light sources are large, square, light-diffusing glass baffles etched in patterns imitating DNA fingerprints. *(38,39, 40,41,42)*

One entire floor of Beckman Laboratory is to be devoted to the study of learning and memory, with emphasis on research on the genes controlling these pro-

(40) Plimpton Seminar Room in Beckman Laboratory.

Talks are presented here by seminar speakers visiting the summer neurobiology training courses conducted in the building. The room is also used by various departments of the Laboratory for large meetings.

(41) Lab bench in Beckman Laboratory.

A chemical hood is visible on the left in this photograph of a typical laboratory for research in neurobiology.

(42) Meeting area of Director's Office in Beckman Laboratory.

Two DNA fingerprint chandeliers are visible in this second floor office in the seminar wing. These were designed by the building's architects to spotlight one of the new DNA techniques widely used today in molecular genetics and forensic medicine.

cesses in the fruit fly *Drosophila*. After a baseline model for understanding these processes has been achieved using the fruit fly, then this model can be tested with higher organisms such as the mouse and, eventually, man. Research and teaching on the other floor will focus on signal transduction processes within the cells of the nervous system of mammals; breakdowns in these lines of communication may lead to diseases such as Alzheimer's and Parkinson's. Work on these two upper floors will benefit from the

presence of the W.M. Keck Structural Biology Laboratory, which occupies the entire lower level. Here all of the facilities at Cold Spring Harbor for the structural analysis of DNA and proteins are to be integrated. The research groups headquartered in the Keck laboratories in part will be studying the flow of genetic information between DNA and proteins and analyzing the protein-DNA interactions at the level of individual atoms.

DOLAN HALL
Centerbrook
1991

Like Beckman Laboratory, Dolan Hall has a splayed footprint and is likewise entered where its two wings come together in the center at the south end of the complex. Here one can enter either the courtyard of the Neuroscience Center at its southern end on foot via a broad passageway underneath the second floor of Dolan Hall or the two residential wings of the building itself whose entrances open onto this covered passageway. The majority of the accommodations are grouped together as suites of two bedrooms (sharing a single bath); a "house-mother's" apartment is located immediately inside the entrance to the east wing. *(43,44)*

Another pedestrian entrance to the Neuroscience Center is located on its west side adjacent to the parking area and between the north end of Dolan Hall and the south end of Beckman Laboratory, next to Hazen Tower. With its multiplicity of

entranceways and passageways the Neuroscience Center has a romantic feeling as well as a classical appearance. The paved courtyard that anchors the buildings of the Center to the site and unites them along its edges has a formal appearance yet there is an intimate neo-Collegiate Gothic air about it. One can almost imagine the copper-inlaid brick stanchions between the windows on Beckman Laboratory to be buttresses not dissimilar to the kind beloved by medieval architects. What is perhaps more difficult to imagine is that below the serene courtyard, which is carpeted with brick pavers and decorated by a single Zelkova tree, is a cavernous mechanical room with boilers, chillers, steam generators, and the like. Even the chimneys for the underground furnaces of the Neuroscience Center are hidden, concealed in the four piers of Hazen Tower.

(43) Dolan Hall from the north.

All of the rooms are singles and share a bath with one other room en suite. Visitor stays range from three days (during professional meetings) to three weeks (for training courses). Most of the bedrooms have either a courtyard or harbor view. In the court-yard in the foreground are three brick carpets, including one planted with a Zelkova tree.

(44) Typical bedroom in Dolan Hall.

The watercolor on the wall depicting Wawepex Building and Jones Laboratory was painted by Centerbrook architect William Grover who convened a special charette (colleagues, family, and friends were invited to participate) for the purpose of rendering projected slide views of Cold Spring Harbor buildings into art for the walls of the Laboratory's newest residence hall.

HAZEN TOWER
Centerbrook
1990

Hazen Tower disguises the Neuroscience Center chimneys the same way that the mosaic-encrusted Tiffany Tower (still standing next door to the Laurel Hollow public beach) camouflaged the smokestack of the heating plant of Louis Comfort Tiffany's Laurelton Hall. Also like the Tiffany Tower, Hazen Tower serves almost as a beacon or landmark on the campus landscape. Standing 65-feet tall just off to the side of the main entrance to the Neuroscience Center courtyard from the parking area, Hazen Tower moreover serves to orient the visitor in time as well as space, for it was built sturdy enough to support a big bell to ring the hours between eight in the morning and eight at night. Weighing nearly a ton, the bronze bell that now hangs in the tower was cast many years ago by the Meneely Foundry of Troy, New York. Recently reconditioned, the bell measures 42 inches in diameter and sounds a deep "G" note. *(45)*

At the top of its lower section the bell tower is joined to a stairway exit from the second floor of Beckman Laboratory. Above this point a helical staircase winds its way up to a circular viewing platform just underneath the bell. Above the bell the four piers of the tower come together to form its top, each of the four sides being adorned with a single polished granite plaque inscribed in gold with the symbol of one of the four nucleotides (**a**denine, **g**uanine, **c**ytosine, and **t**yrosine) that compose every molecule of DNA.

(45) Hazen Tower from the roof of Beckman Laboratory.

Each face of this 65-foot bell tower bears a granite plaque inscribed with the initial of one of the four nucleotides that make up the backbone of the DNA molecule: **c**ytocine, **g**uanine, **a**denine, and **t**yrosine. Inside hangs a 42-inch cast bronze bell that sounds a deep "G" note. (The New Cabins can be seen in the background on the right.)

Site planning and landscaping come of age at the
Neuroscience Center

Due of course to its height, Hazen Tower is the most visible part of the Neuroscience Center as viewed from across the harbor. This distant view was an important consideration not only for the architects from Essex, but also for the landscape architects Keith Simpson Associates, who earlier had furnished the landscape plan for Grace Auditorium and the hillside next to it that leads up to the Neuroscience Center site. The views of Beckman Laboratory and Dolan Hall are filtered through a hillside forest canopy that was discreetly thinned out to provide harbor vistas from the main seminar areas without dissolving the illusion that the Center is still surrounded by woods. The decision early on to save space by stacking the necessary parking areas on two levels at the rear of the site, which was on a steep incline, had required the erection of massive retaining walls as an essential part of the first phase of the construction project, followed immediately by pouring the concrete for the parking garage itself.

The landscaping for the disturbed hillside that resulted behind the parking garage (between it and the New Cabins to the west) was similar to that specified in 1986 for the hillside to the west of Grace Auditorium—native trees randomly placed as though they had grown up there. Once again more formal and ornamental plantings were installed to define the edges of the parking area itself, including a long row of Callery pear trees, the same attractive flowering trees that were planted along Bungtown Road next to Grace Auditorium.

Cold Spring Harbor Laboratory is becoming a
"university of DNA"

Beckman Laboratory and Dolan Hall (together with Hazen Tower), nearing completion as Cold Spring Harbor Laboratory's Centennial drew to a close in 1990, are the newest additions to the list of structures at or near Cold Spring Harbor frequented by visitors with a scientific bent. The buildings in this collection, in which Grace Auditorium and Blackford Hall on Bungtown Road and the Meeting House and Sammis Hall at Banbury Center in Lloyd Harbor also figure prominently, are from varying vintages, a mixture of the old with the new, all in their current guises the products of designs emanating from the drawing boards of the Essex, Connecticut, architectural firm now known as Centerbrook. Collectively these high-visibility buildings typify all three of the ongoing aims of the science at Cold Spring Harbor—research, meetings, and teaching—and also a 100-year-old mandate to house as many of the participants in these activities as possible on-grounds, for the sake of the science itself.

(46) Dolan Hall at dusk.

Night falls on Cold Spring Harbor's "university of DNA."

Two of these elements were present from the founding of the Biological Laboratory in 1890. In those early days the scientists occupied their working hours with courses and research in a building especially erected for them by John D. Jones in 1893. This Colonial-style "school-house and laboratory" symbolized the dual purposes of the Bio Lab and it was the first real "house for science" at Cold Spring Harbor. For living accommodations for the summer biologists there were old homes in the immediate vicinity that formerly had belonged to other members of Cold Spring Harbor's first family, the Joneses, and dated back to the whaling days when the western shore of the harbor was known as Bungtown—more "houses for science."

Later research buildings erected after the turn of the century imitated European styles, especially the Italian, and even a famous laboratory built on the shores of the Bay of Naples. Then came a period in the late 1920s of building laboratories at Cold Spring Harbor that looked just like the old houses in the area. The annual Symposium on Quantitative Biology was inaugurated shortly thereafter in 1933. As a result summer visitors arrived in increasing numbers and many found themselves staying in Cabins or a Motel. With the post–World War II explosion of biology in the 1950s, newer and larger scientific facilities appeared along Bungtown Road, but in many ways it still looked like a "summer camp for biologists."

As the older facilities became spruced up, enlarged, and winterized starting in the 1970s, Bungtown became a bustling year-round "village of science." Today it is still a single-industry town, its inhabitants all engaged in state-of-the-art molecular genetics, whether attending a meeting, pursuing research, or taking or teaching a course. Newly enhanced conference and housing facilities now allow them to pursue these activities with single-minded dedication in comfortable surroundings. Finally, with the addition of the major new structures that comprise the Neuroscience Center, "houses for science" has taken on a newer, more permanent connotation. Bungtown is rapidly becoming a university town—"DNA Town." (46)

APPENDICES

CHARLES B. DAVENPORT REGINALD G. HARRIS ALBERT F. BLAKESLEE

MILISLAV DEMEREC JOHN CAIRNS JAMES D. WATSON

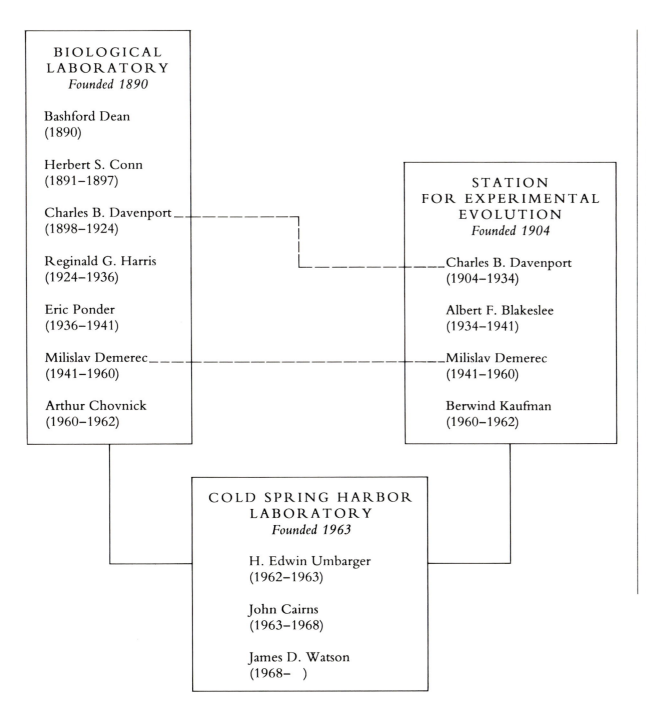

BIOLOGICAL
LABORATORY
Founded 1890

Bashford Dean
(1890)

Herbert S. Conn
(1891–1897)

Charles B. Davenport
(1898–1924)

Reginald G. Harris
(1924–1936)

Eric Ponder
(1936–1941)

Milislav Demerec
(1941–1960)

Arthur Chovnick
(1960–1962)

STATION
FOR EXPERIMENTAL
EVOLUTION
Founded 1904

Charles B. Davenport
(1904–1934)

Albert F. Blakeslee
(1934–1941)

Milislav Demerec
(1941–1960)

Berwind Kaufman
(1960–1962)

COLD SPRING HARBOR
LABORATORY
Founded 1963

H. Edwin Umbarger
(1962–1963)

John Cairns
(1963–1968)

James D. Watson
(1968–   )

The Biological Laboratory, founded at Cold Spring Harbor in 1890, was originally under the auspices of the Brooklyn Institute of Arts and Sciences; later its management was assumed by the Long Island Biological Association, incorporated in 1924. The Station for Experimental Evolution, founded at Cold Spring Harbor in 1904 by the Carnegie Institution of Washington, was reorganized in 1921 as the Department of Genetics under the same operating body. The facilities managed by both the Long Island Biological Association and the Carnegie Institution of Washington were amalgamated in 1962 as the Cold Spring Harbor Laboratory of Quantitative Biology, which in 1970 was shortened to Cold Spring Harbor Laboratory. Photographs of directors who have had a significant impact on the history of the Laboratory are shown on the facing page and their terms of office are given in the organizational chart.

At the first meeting, held July 1, 1933, Dr. Harris made the following remarks:

"In opening this conference it seems desirable to state briefly why the Biological Laboratory at Cold Spring Harbor has invited you to join together in experimental work and in conference this summer.

"The officers of the Laboratory are interested in the development of an institute in which biologists, chemists, physicists and mathematicians will cooperate in the future opening, and beneficial use, of the vast territory of quantitative biology.

"The initial step toward the accomplishment of this aim was taken, in respect to the all-year work of the Laboratory, in 1928, when Dr. Hugo Fricke, a physicist, was appointed to establish and direct our laboratory for biophysics. The second move has brought into being the conference which now begins.

"The present meeting is the inauguration of a plan whereby each summer a group of mathematicians, physicists, chemists and biologists, actively interested in a specific aspect of quantitative biology, or in methods and theories applicable to it, will be invited to carry on their work, to give lectures and to take part in symposia at the Laboratory. A given group in residence here will necessarily be relatively small, but members of the group will be chosen with the aim that every important aspect of a particular subject is adequately represented from the physical and chemical, as well as from the biological point of view; and that the whole span of a subject, from theories of physics to application to medicine, is covered.

"Workers, other than those forming the group in residence during a summer, are invited to take part in the symposia in connection with the group meeting.

"It is expected that many advantages will be secured through the operation of the plan. Outstanding among these is the value of the meetings to the men who form the group.

"Another advantage lies in the fact that the presence of such a group at Cold Spring Harbor each summer will aid the Laboratory in its primary aim of being of as much service as it can be in advancing biology, and of its corollary aim of fostering a closer relationship between the basic sciences and biology.

"This is in thorough consonance with the historical aims of this and other, partially or wholly summer laboratories, namely that they should be centers of growth and dissemination of new methods and ideas in biology. In order to make available to all workers the methods and ideas which will be set forth in the group meetings, from year to year, we plan to publish, as a monograph, the lectures and essential parts of the discussion.

"May I ask you, then, in your lectures, to give special consideration to theoretical and controversial aspects, that the discussion may be both significant and creative, and that those conferences may be of the greatest possible value not only to those of us who take part in them, but also to those who will have occasion to refer to them."

I   (1933)   *Surface Phenomena*
             30 participants

II (1934)   *Aspects of Growth*
             58 participants

III (1935)   *Photochemical Reactions*
             57 participants

IV (1936)   *Excitation Phenomena*
             54 participants

V   (1937)   *Internal Secretions*
             79 participants

VI (1938)   *Protein Chemistry*
             71 participants

VII (1939)   *Biological Oxidations*
             58 participants

VIII (1940) *Permeability and the Nature of Cell Membranes*
             41 participants

IX   (1941) *Genes and Chromosomes: Structure and Organization*
             80 participants

X   (1942) *The Relation of Hormones to Development*
             117 participants

XI   (1946) *Heredity and Variation in Microorganisms*
             136 participants

XII (1947) *Nucleic Acids and Nucleoproteins*
             165 participants

XIII (1948) *Biological Applications of Tracer Elements*
             163 participants

XIV (1949) *Amino Acids and Proteins*
             186 participants

XV   (1950) *Origin and Evolution of Man*
             129 participants

XVI (1951) *Genes and Mutations*
             305 participants

XVII	(1952)	*The Neuron* 125 participants
XVIII	(1953)	*Viruses* 272 participants
XIX	(1954)	*The Mammalian Fetus: Physiological Aspects of Development* 136 participants
XX	(1955)	*Population Genetics: The Nature and Causes of Genetic Variability in Population* 191 participants
XXI	(1956)	*Genetic Mechanisms: Structure and Function* 281 participants
XXII	(1957)	*Population Studies: Animal Ecology and Demography* 134 participants
XXIII	(1958)	*Exchange of Genetic Material: Mechanism and Consequences* 248 participants
XXIV	(1959)	*Genetics and Twentieth Century Darwinism* 196 participants
XXV	(1960)	*Biological Clocks* 150 participants
XXVI	(1961)	*Cellular Regulatory Mechanisms* 244 participants
XXVII	(1962)	*Basic Mechanisms in Animal Virus Biology* 218 participants
XXVIII	(1963)	*Synthesis and Structure of Macromolecules* 288 participants
XXIX	(1964)	*Human Genetics* 143 participants
XXX	(1965)	*Sensory Receptors* 161 participants
XXXI	(1966)	*The Genetic Code* 349 participants
XXXII	(1967)	*Antibodies* 286 participants
XXXIII	(1968)	*Replication of DNA in Microorganisms* 354 participants
XXXIV	(1969)	*The Mechanism of Protein Synthesis* 287 participants
XXXV	(1970)	*Transcription of Genetic Material* 333 participants
XXXVI	(1971)	*Structure and Function of Proteins at the Three-dimensional Level* 193 participants
XXXVII	(1972)	*The Mechanism of Muscle Contraction* 198 participants
XXXVIII	(1973)	*Chromosome Structure and Function* 280 participants
XXXIX	(1974)	*Tumor Viruses* 346 participants
XL	(1975)	*The Synapse* 244 participants
XLI	(1976)	*Origins of Lymphocyte Diversity* 326 participants
XLII	(1977)	*Chromatin* 375 participants
XLIII	(1978)	*DNA: Replication and Recombination* 428 participants
XLIV	(1979)	*Viral Oncogenes* 411 participants
XLV	(1980)	*Movable Genetic Elements* 302 participants
XLVI	(1981)	*Organization of the Cytoplasm* 234 participants
XLVII	(1982)	*Structures of DNA* 244 participants
XLVIII	(1983)	*Molecular Neurobiology* 223 participants
XLIX	(1984)	*Recombination at the DNA Level* 248 participants
L	(1985)	*Molecular Biology of Development* 275 participants
LI	(1986)	*Molecular Biology of Homo sapiens* 278 participants
LII	(1987)	*Evolution of Catalytic Function* 187 participants
LIII	(1988)	*Molecular Biology of Signal Transduction* 430 participants
LIV	(1989)	*Immunological Recognition* 446 participants
LV	(1990)	*The Brain* 281 participants

The definitions in this glossary are excerpted from Henry H. Saylor's *Dictionary of Architecture*. Saylor designed three laboratories at Cold Spring Harbor in the mid to late 1920s, including Nichols Building which is used here to illustrate the term "transom." All of the architectural features depicted are found on one or more of the buildings discussed in the text.

Arcade a range of arches with their supports

Baluster a...form of upright which, in series [balustrade], supports a handrail

Bay window a window or windows in a wall that projects angularly from another wall and from the ground up

ARCADE (Carnegie Library)

Excerpts from *Dictionary of Architecture* by Henry H. Saylor, copyright 1952 John Wiley & Sons, Inc. Reprinted by permission of John Wiley & Sons.

BALUSTER (Cole Cottage)

BAY WINDOW (Airslie)

BOARD & BATTEN (Davenport House)

BULL'S-EYE (DNA Learning Center)

BRACKET (Davenport House)

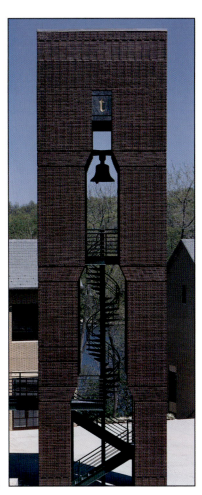

CAMPANILE (Hazen Tower)

Board & batten [construction utilizing] flat pieces of wood [boards] and narrow strip members [battens]...covering the vertical joint between [the] boards

Bracket a supporting member for a projecting floor... sometimes in the shape of an inverted "L"

Bull's-eye a circular window or louver, an oeil-de-boeuf

Campanile a bell tower

Cantilever a form of construction in which a beam or series of beams is supported by a downward force behind a fulcrum

Capital the top member...of a column...or pilaster

Clapboard a board that is thin on one edge and thicker on the other, to facilitate overlapping horizontally to form a weatherproof, exterior wall surface

Cupola a terminal structure, square to round in plan, rising above a main roof

Dentil one of a series of block-like projections forming a molding

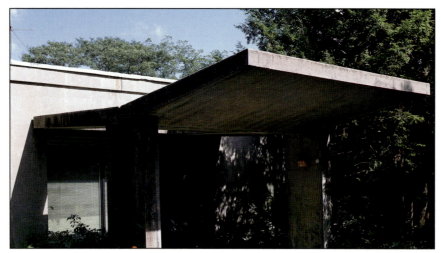

CANTILEVER (Demerec Laboratory)

CAPITAL (Carnegie Library)

CLAPBOARD (Davenport House)

DENTIL (Jones Laboratory)

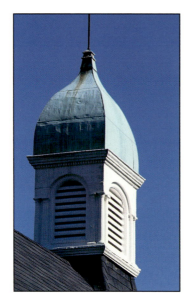

CUPOLA (Jones Laboratory)

DORMER (Williams House)

Dormer a minor gable in a pitched roof, usually bearing a window or windows on its front vertical face

Fanlight an overdoor window, semi-elliptical or semicircular in shape, with radial muntins or leads

Finial a terminal form at the top of a spire, gable,... pinnacle, or other point of relative height.

Gable the upper part of a terminal wall under the ridge of a pitched roof

FANLIGHT (Robertson House)

FINIAL (Davenport House)

GABLE (Nichols Building)

Gambrel a form of roof in which the angle of pitch is abruptly changed between ridge and eaves

Half-timber descriptive of sixteenth- and seventeenth-century buildings, particularly in England and France, where the interstices in the framework of heavy timbers were filled with brick or plaster

Keystone the wedge-shaped top member of an arch

Modillion a bracket form used in a series under a corona [the lowest member of a classical cornice]

GAMBREL (Airslie)

HALF-TIMBER (Davenport House)

KEYSTONE (McClintock Laboratory)

MODILLION (Jones Laboratory)

MONITOR (Carnegie Library)

Monitor a continuous section of roof raised to admit light on a vertical plane

Parapet a low, retaining wall at the edge of a roof, porch, or terrace

Pediment the triangular face of a roof gable, especially in its classical form

Pent roof a roof of a single sloping plane

PARAPET (McClintock Laboratory)

PENT ROOF (Cole Cottage)

PEDIMENT (Robertson House)

Pilaster an engaged pier of shallow depth

Porte-cochère a shelter for vehicles outside an entrance doorway

Portico an entrance porch

Quoin one of the corner stones of a wall when these are emphasized by size...or by a difference in texture

PILASTER (Airslie)

PORTE-COCHERE (Olney House)

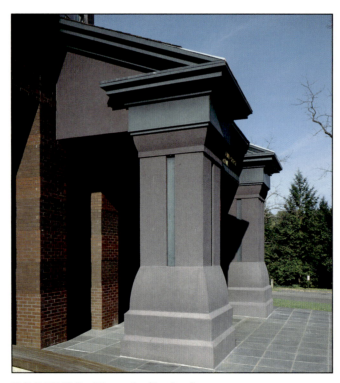

PORTICO (Grace Auditorium)

QUOIN (McClintock Laboratory)

340

ROUNDEL (Carnegie Library)

RANDOM RUBBLE (Jones Laboratory)

Roundel a small circular window or panel

Random rubble untrimmed stone laid with no course lines

Shed roof a roof having only one sloping plane

Shingle a wedge-shaped piece of wood as used in overlapping courses to cover a roof or an outside wall surface

SHED ROOF (James Laboratory)

SHINGLE (Olney House)

Side light one of a pair of narrow windows flanking a door

Spandrel the surface at the side of a half-arch between a vertical line at the bottom of the archivolt and a horizontal line through its top

Transom an opening over a door or window...containing a glazed or solid sash

Vergeboard (bargeboard) the vertical-face board following and set back under the roof edge of a gable, sometimes decorated by carving

SIDE LIGHT (Delbrück Laboratory)

SPANDREL (Carnegie Library)

TRANSOM (Nichols Building)

VERGEBOARD (Firehouse)

ARCHITECTURE

AIA Guide to Long Island Architecture: Three Centuries of Architecture in Nassau and Suffolk Counties. Garden City, New York: Dover Publications, in press.
~ A joint production of the Long Island Chapter of the American Institute of Architects and the Society for the Preservation of Long Island Antiquities, copiously illustrated.

Beveridge, Charles E. and Carolyn F. Hoffman, with Shary Page Berg and Arleyn A. Levee. *The Master List of Design Projects of the Olmsted Firm 1857–1950.* Boston: Massachusetts Association for Olmsted Parks, 1987.
~ Lists by town or village all of the Long Island commissions (among others) of this celebrated landscape architectural firm headquartered in Brookline, Massachusetts.

Blumenson, John J.-G. *Identifying American Architecture: A Pictorial Guide to Styles and Terms, 1600–1945.* Nashville: American Association for State and Local History, 1979.
~ Illustrates 39 styles, using photographs overlaid with numbers referring to the architectural hallmarks of each style as shown in the photos. There is also a similarly illustrated Pictorial Glossary and a selected Bibliography that provides an excellent introduction to the literature of American architectural history.

Bullock, Orin M., Jr. *The Restoration Manual: An Illustrated Guide to the Preservation and Restoration of Old Buildings.* New York: Van Nostrand Reinhold Company, 1983. (First published in 1966 by Silvermine Publishers.)
~ The cover of the 1983 paperback edition of this preservation classic (sponsored by the American Institute of Architects) features a photograph of Davenport House taken shortly after its 1980 restoration.

Exterior Decoration: Victorian Colors for Victorian Homes. Philadelphia: The Atheneum of Philadelphia, 1976. (Facsimile edition of book originally published in 1885 by the Devoe Paint Manufacturing Company.)
~ Circa 1885 Olney House on the grounds of Cold Spring Harbor Laboratory was repainted in shades of green and olive based on the recommendations contained in Plate II which illustrates an alternative mode of painting "a private residence in the vicinity of New York, built in the modern 'Queen Anne' style of architecture." Plate I, another scheme for painting a Queen Anne-style house, is remarkably similar to the original exterior decoration scheme of Davenport House (as determined by an historic paint color analysis and recently restored) that featured golden and ochre tones.

Gandee, Charles K. "Preserving the Quietude." *Architectural Record*, April 1982, pp. 95–103.
~ Showcases the early work at Cold Spring Harbor Laboratory by the Essex, Connecticut, firm founded by Charles W. Moore.

"Gleaming Laboratory Cubes in a Rustic 19th Century Shell." *AIA Journal*, Mid-May 1981, pp. 252–253.
~ Commemorates Jones Laboratory and its winning a 1981 Honor Award from the American Institute of Architects in the Continued Use category.

Hewitt, Mark Alan. *The Architect and the American Country House, 1890–1940.* New Haven: Yale University Press, 1990.
~ Some of the leading domestic architects of the heyday of the American country house also designed laboratories at Cold Spring Harbor; this handsome big book includes informative Architects' Biographies at the back.

Jensen, Robert and Patricia Conway. *Ornamentalism: The New Decorativeness in Architecture and Design.* New York: Clarkson N. Potter, Inc., 1982.
~ A detail of Sammis Hall at Banbury Center is the cover photograph of this profusely illustrated showcase of Post-Modern Classic design.

Krieg, Joann P. *Long Island Architecture.* Interlaken, New York: Heart of the Lakes Publishing, 1991.
~ A selection of papers presented at the conference sponsored by the Long Island Studies Institute (Hofstra University, Hempstead, Long Island, New York) in June of 1989 on Building Long Island: Architecture & Design; Tools & Trade. Topics range from seventeenth century Dutch buildings to Modernist beach houses. Contains an excellent Bibliography on Long Island architecture by Long Island Studies Institute director Natalie A. Naylor.

MacKay, Robert B. and Carol Traynor. *Long Island Country Houses and Their Architects, 1860–1940.* New York: W.W. Norton, in press.
~ All the architectural firms active during the mansion period on Long Island have merited individual chapters in this long-awaited book.

Moore, Charles W., Gerald Allen and Donlyn Lyndon. *The Place of Houses.* New York: Holt, Rinehart and Winston, 1974.
~ Lucidly written and profusely illustrated with photographs and drawings by the authors, this enjoyable and widely quoted book explains how houses take shape in their designers' eyes.

Randall, Monica. *The Mansions of Long Island's Gold Coast.* New York: Hastings House Publishers, 1979.
~ Photographs galore (old and new).

Saylor, Henry H. *Dictionary of Architecture.* New York: John Wiley & Sons, 1952.
~ Compiled by the first full-time editor of the *AIA Journal,* who was also the architect of both Davenport and James Laboratories and Nichols Building at Cold Spring Harbor.

Sclare, Liisa and Donald Sclare. *Beaux-Arts Estates: A Guide to the Architecture of Long Island.* New York: The Viking Press, 1980.
~ This survey of 32 estates on Long Island was the first book to treat in a comprehensive and serious manner the stately homes here and the architects who designed them, including some of New York's most talented practitioners.

Speaking a New Classicism: American Architecture Now. With essays by Guest Curator, Helen Searing and Henry Hope Reed. Northampton, Massachusetts: Smith College Museum of Art, 1981.
~ An architectural drawing of Sammis Hall at Banbury Center is the cover illustration to this catalog of the exhibit on Post-Modern Classicism that was mounted at the Smith College Museum of Art in the spring of 1981.

Stern, Robert A.M. *Pride of Place: Building the American Dream.* Boston: Houghton Mifflin Company and New York: American Heritage, 1986.
~ Cold Spring Harbor Laboratory is featured in the chapter on "Academical Villages: The Places Apart" in this volume that accompanied the Mobil Oil Company-sponsored Public Broadcasting System's television series of the same name which included some dramatic footage shot at dawn at Cold Spring Harbor.

Whiffen, Marcus. *American Architecture Since 1780: A Guide to the Styles*. Cambridge, Massachusetts: The M.I.T. Press, 1976.
~ Forty styles are illustrated by photographs; includes Bibliography and Glossary.

LABORATORY PUBLICATIONS

Annual reports: Biological Laboratory, 1890–1963; Carnegie Institution of Washington Station for Experimental Evolution, 1904–1920; Carnegie Institution of Washington Department of Genetics, 1921–1963; Cold Spring Harbor Laboratory (of Quantitative Biology), 1964–.

Booklets: Dedication—Cabins, 1989; Delbrück Laboratory, 1981; DNA Learning Center, 1988; Grace Auditorium, 1986; Harris Building, 1982; Hershey Building, 1979; Neuroscience Center, 1991; Page Laboratory, 1987; Sambrook Laboratory, 1985; Sammis Hall, 1981. Other—Davenport's Daughter (art exhibit), 1990; Nothing But Steel (sculpture exhibit), 1987.

Cold Spring Harbor Laboratory and Long Island: Partners for the Future (Centennial Journal), 1990.

The First Hundred Years: A History of Man and Science at Cold Spring Harbor (illustrated historical booklet prepared for the Second Century Campaign by David Micklos with Susan Zehl, Daniel Schechter, and Ellen Skaggs), 1988.

Harbor Transcript (Cold Spring Harbor Laboratory newsletter, Susan Cooper, editor), 1988–.

Publications Catalog of Cold Spring Harbor Laboratory Press, 1990–1991.

LOCAL HISTORY

Atlas of Nassau County. E. Belcher Hyde, 1898.
~ The western side of Cold Spring Harbor is designated on this map as "Bungtown."

Burlingham, Michael John. *The Last Tiffany: A Biography of Dorothy Tiffany Burlingham*. New York: Atheneum, 1989.
~ Insights into the life and times of the Tiffanys and also their neighbors the de Forests. Both families owned homes near the Laboratory.

Cold Spring Harbor Soundings. Cold Spring Harbor, New York: Cold Spring Harbor Village Improvement Society, 1953.
~ A compilation of separately authored institutional histories of all of the community organizations and businesses at Cold Spring Harbor, past and present.

Earle, Walter K. *Out of the Wilderness: Being an Account of the Settlement of Cold Spring Harbor, Long Island, and the Activities of Some of the Settlers from the Beginning to the Civil War*. Cold Spring Harbor, New York: The Whaling Museum, Inc., 1966.
~ Earle coined the phase "Jones Industries" to collectively describe all the businesses the sons of John Jones were involved in at Cold Spring Harbor in the first half of the nineteenth century.

Jones, John H. *The Jones Family of Long Island*. New York: Tobias A. Wright, 1907.
~ John Henry Jones was a nephew of Major William Jones, who built the house now called Airslie on the grounds of Cold Spring Harbor Laboratory.

MacKay, John, editor. *"Walls Have Tongues": Oyster Bay Buildings And Their Stories*. Oyster Bay, New York: Oyster Bay Historical Society, 1977.
~ Each building is illustrated by an attractive full-page drawing.

MacKay, Robert B., Geoffrey L. Rossano, and Carol A. Traynor, editors. *Between Ocean and Empire: An Illustrated History of Long Island*. Northridge, California: Windsor Publications, Inc., in cooperation with the Society for the Preservation of Long Island Antiquities and the Long Island Association, 1985.
~ An attractive and informative book for anyone curious about the Island's past.

Murphy, Robert Cushman. *Fish-Shaped Paumanok: Nature and Man on Long Island*. Great Falls, Virginia: Waterline Books, 1991. (Reprint, together with Foreword by Steven C. Englebright, of book first published at Philadelphia in 1964 by the American Philosophical Society.)
~ The author, a native and lifelong resident of Long Island, headed the Department of Ornithology at the American Museum of Natural History as Lamont Curator of Birds. His major work was *Oceanic Birds of South America* (1936). He also wrote *Logbook for Grace* (1947), a volume of letters to the bride he left at home when he sailed off in 1912 on a square-rigged Yankee whaler on his first long scientific journey to the South Atlantic. He was a founding member of the Whaling Museum Society in Cold Spring Harbor and a long-term director of the Long Island Biological Association.

Stein, Jean and George Plimpton. *Edie: An American Biography*. New York: Alfred A. Knopf, 1982.
~ "Edie" was a de Forest and a protégée of Andy Warhol.

Stone, Gaynell. "Long Island as America: A New Look at the First Inhabitants." *Long Island Historical Journal* 1 (Spring 1989.)
~ Insights into the Native Americans who first inhabited Cold Spring Harbor.

Tooker, William Wallace. *The Indian Place-Names on Long Island and Islands Adjacent, With Their Probable Significations*. (Edited, with an Introduction by Alexander F. Chamberlain, Ph.D.) Port Washington, New York: Ira J. Friedman Division, Kennikat Press, 1975.
~ Gives the derivation of "Wawepex," the name for the Native American settlement on the western shore of Cold Spring Harbor.

SCIENCE

Allen, Garland E. "The Eugenics Record Office at Cold Spring Harbor, 1910–1940: An Essay in Institutional History," in *Osiris: A Research Journal Devoted to the History of Science and Its Cultural Influences*, 1986 (Second Series, Volume 2), pp. 225–264.
~ Describes the fate of the eugenics records compiled at Cold Spring Harbor between 1910 and 1940.

Allen, Garland E. *Thomas Hunt Morgan*. Princeton: Princeton University Press, 1978.
~ Laboratory director Charles Davenport figures prominently in the section of Chapter VI entitled "Morgan, Eugenics, and Social Biology," starting on p. 227.

Cattell, J. McKeen and Jacques Cattell. *American Men (and Women) of Science: A Biographical Directory*. New York: The Scientific Press and Bowker Publications, 1906–1990 (17 Editions).
~ The first 8 editions (1906, 1910, 1921, 1927, 1933, 1938, 1944, and 1949) were single-volume editions; multiple-volume editions began with the 9th Edition starting in 1955, the words "and Women" being added to the title with the inception of the 12th Edition in 1971.

Glass, Bentley. *A Guide to the Genetics Collections of the American Philosophical Society* (American Philosophical Society Library Publication Number 13.) Philadelphia: American Philosophical Society, 1988.
~ Contains references to the papers of Laboratory scientists A.F. Blakeslee, C.B. Davenport, M. Demerec, G.H. Shull, C.B. Bridges, B.P. Kaufmann and others.

Glass, Bentley. *A History of Genetics at Cold Spring Harbor*. Cold Spring Harbor, New York: Cold Spring Harbor Laboratory Press, forthcoming.
~ The author was a frequent summer visitor and Symposium participant at Cold Spring Harbor in the 1940s and 1950s while a professor at Goucher College and later Johns Hopkins University. After coming to the State University of New York at Stony Brook in the mid-1960s he served as chairman of the Board of Trustees of Cold Spring Harbor Laboratory from 1967 to 1973.

Keller, Evelyn Fox. *A Feeling for the Organism: The Life and Work of Barbara McClintock*. New York: W.H. Freeman and Company, 1983.
~ Published the year McClintock won her Nobel prize.

Kevles, Daniel J. *In the Name of Eugenics: Genetics and the Uses of Human Heredity*. New York: Alfred A. Knopf, 1985.
~ Contains a deftly drawn portrait of the Laboratory's early and influential director—Chapter III, "Charles Davenport and the Worship of Great Concepts" (pp. 41–56)—and a valuable "Essay on Sources."

Kittredge, Mary. *Barbara McClintock*. New York and Philadelphia: Chelsea House Publishers, 1991.
~ A volume in the series *American Women of Achievement* written for young adults and introducing "50 women whose actions, ideas and artistry have helped shape the course of American history;" introductory essay (to series) by Matina S. Horner, President Emerita of Radcliffe College.

MacDowell, E. Carleton. "Charles Benedict Davenport, 1866–1944: A Study of Conflicting Influences." *Bios*, Vol. 1, No.1, 1946, pp. 1–50.
~ A revealing portrait by one of Charles Davenport's Cold Spring Harbor protégés. Includes a "Bibliography of Charles B. Davenport," 1890–1945.

Paul, Diane B. and Barbara A. Kimmelman. "Mendel in America: Theory and Practice, 1900–1919," in: Rainger, Ronald, Keith R. Benson, and Jane Maienschein, editors, *The American Development of Biology* (Philadelphia: University of Pennsylvania Press, 1988), pp. 281–310.

Micklos, David A. and Greg A. Freyer. *DNA Science: A First Course in Recombinant DNA Technology*. Cold Spring Harbor, New York: Cold Spring Harbor Laboratory Press and Carolina Biological Supply Company, 1990.
~ A textbook and laboratory manual for high school and junior college students.

Stent, Gunther S. *The Coming of the Golden Age: A View of the End of Progress*. Garden City, New York: The Natural History Press, 1969.
~ Covers the role of Cold Spring Harbor in the birth of molecular biology.

Watson, James D. *The Double Helix*. New York: Atheneum Press, 1968.
~ A personal account of the 1953 discovery of the double helical nature of the structure of DNA (deoxyribonucleic acid).

Watson, James D., Nancy D. Hopkins, Jeffrey W. Roberts, Joan Argetsinger Steitz, and Alan Weiner. *Molecular Biology of the Gene*. Menlo Park, California: Freeman Publishers, 1989. (Revised edition of 3rd ed. [1976])
~ Authoritative genetics text, first published in 1965.

Watson, James D., John Tooze, and David T. Kurtz. *Recombinant DNA: A Short Course*. New York: W.H. Freeman and Company (Scientific American Books), 1983. (Second edition in press; authors are James D. Watson, Michael Gilman, Jan Witkowski, and Mark Zoller.)
~ Contains a "Recombinant DNA Dateline" (1971–1983)

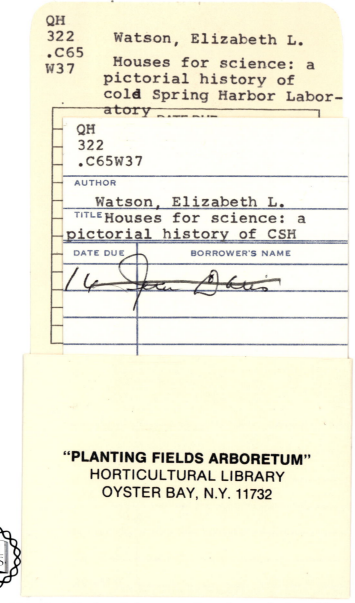

The Cold Spring Harbor Laboratory Press logo, designed by
James Suddaby in 1989, is an adaptation of the Cold Spring
Harbor Laboratory logo which consists of the letters CSH
circumscribed by a closed circular double helix, a representation of an SV40 DNA molecule. The Laboratory logo was
adopted in 1970 shortly after James D. Watson was appointed director.

Typeset by Cold Spring Harbor Laboratory Press, Cold Spring Harbor, New York
in Bembo from the Monotype® Typeface Library of Adobe Systems Incorporated
Color separations and printing by Nimrod Press, Printers and Engravers, Boston, Massachusetts
Printed on 100-pound Celesta Litho Dull paper by Westvaco Corporation
Bound by Bridgeport Bindery, Agawam, Massachusetts, in Brillianta cover cloth by Scholco

Colonial and Federal					Italianate	Victorian	Colonial Revival	Renaissance
1800	1810	1820	1830	1840		1880	1890	1900
Osterhout Cottage ca. 1800 (Acq. 1893; reconst. 1969) Airslie 1806 (Acq. 1943)		Yellow House ca. 1820 (Acq. 1972) Wawepex Building ca. 1825 (Acq. 1893)	Hooper House ca. 1835 (Acq. 1893) Williams House ca. 1835 (Acq. 1926; reconst. 1977)	Stewart House ca. 1840 (First home of Eugenics Record Office 1910-1914; now a private residence)		Davenport House 1884 (Acq. 1903; rest. 1980) Olney House ca. 1885 (Acq. 1973) Fish Hatchery 1887 (Dem. 1950s) (First home of Bio Lab 1890-1893)	Jones Laboratory 1893	Main Building 1905 (Carnegie Library since 1953) Firehouse 1906 (Reloc. to Bio Lab 1930) Blackford Hall 1907

Two centuries of building at Bungtown and environs